The Gwithian Landscape

Molluscs and Archaeology on Cornish Sand Dunes

Thomas M. Walker

With contributions from

Rowena Y. Banerjea and C. Rob Batchelor

Archaeopress Archaeology

ARCHAEOPRESS PUBLISHING LTD
Summertown Pavilion
18-24 Middle Way
Oxford OX2 7LG

www.archaeopress.com

ISBN 978 1 78491 803 3
ISBN 978 1 78491 804 0 (e-Pdf)

Cover: Aerial view of the Godrevy Headland, looking north-east.

Charles Thomas at Godrevy Barrow during excavations in 2012

This work is dedicated to Charles Thomas

1928–2016

Whose studies at Gwithian and elsewhere in Cornwall and throughout Britain have inspired numerous archaeologists.

Without his enthusiastic support this study could not have been undertaken.

Contents

List of figures

List of tables

Preface

Coastal sand dunes are common in Britain, especially along western Atlantic coasts, and archaeological sites are well preserved under blown sand. One of the best excavated sites is at Gwithian, on the east side of St Ives Bay on the north coast of Cornwall. This is a multiperiod site, with evidence of human activity extending from the Mesolithic to the post-medieval period. Major excavations at Gwithian took place in the 1950s and 1960s under the direction of Charles Thomas, to whom this work is dedicated; these revealed a Bronze Age settlement and a post-Roman industrial area in the valley of the Red River and a medieval manor house on the summit of Godrevy Towans. There were repeated phases of occupation and abandonment.

Few studies of British coastal sites have included palaeoenvironmental evidence, and even fewer include securely dated stratigraphy. The present study uses a multiproxy approach to investigate the palaeoenvironment of the Gwithian area, with molluscs being the principal analytical method. Chronology is established by radiocarbon dating and optically stimulated luminescence so that episodic human activity can be related to periods of sand blow and instability. Attempts are made to assess how Holocene palaeoclimate variation influenced the patterns of settlement establishment and abandonment. Evidence is also sought concerning the history of mineral mining in the Red River catchment area.

Twenty three cores and a test pit were examined along a transect from the Red River valley basin to the summit of Godrevy Towans. Molluscan studies in the wider landscape of Gwithian expand understanding of the palaeoenvironment from the late Neolithic up to recent times.

The study establishes that initial sand deposition was about 3000 BC, with further marked periods of sand blow in the early and late Bronze Age, the Iron Age and the mid-medieval period. Some, but not all, sand blow occurrences correlate with periods of settlement occupation and abandonment, and with known palaeoclimate episodes such as the Little Ice Age. Difficulty in establishing periods of sand conflation and deflation and how this may lead to errors in chronology are discussed. There is weak evidence for mining activity in the Bronze Age, but strong evidence from *c.* 1050 AD.

The chronology of some mollusc extinctions and introductions is refined. *Xerocrassa geyeri* did not become extinct at Gwithian until the end of the early Bronze Age and *Cochlicella acuta* was present from the late Neolithic, about 2100 BC.

Foreword

by Jacqueline A. Nowakowski FSA
Principal Archaeologist, Cornwall Archaeological Unit
Cornwall Council, Cornwall

Gwithian, in West Cornwall, is not only a beautiful stretch of the UK coastline, but it is an archaeologist's dream landscape. Broad but expansive undulating blankets of wind-blown sands, today chequered by fields and the dunes, bonded in places by scrubby maritime prickly gorse, safeguard lost places and "lock-in" hidden landscapes. The archaeological fieldwork directed by Charles Thomas at Gwithian for over 50 years introduced the remarkable human stories hidden within this special coastal place to a wider world: for at Gwithian it is possible to uncover lost settlements (ancient and historic), and encounter animal footprints and plough marks preserved in ancient land surfaces long sealed by the sands. The Bronze Age settlements discovered at Gwithian by Charles in particular inspired a generation of landscape archaeologists and kindled a holistic approach to landscape studies. I for one learnt about the remarkable spade and plough marks preserved in the relict surfaces of these ancient fields in far-flung Cornwall as an undergraduate in Sheffield in the late 1970s. Like many, these discoveries inspired me to become a field archaeologist. Years later I happily got the chance to see for myself these unique discoveries when Charles returned to the Bronze Age settlement and re-exposed the deep multi-layered sequence of "occupation" horizons, so that we could take new scientific samples in 2005. They reappeared in fresh sharp outline and pristine condition as if they had only been made yesterday! And when Tom Walker and his team returned to Gwithian in 2012 to carry out his own small excavation (reported on here in Chapter X), I again was fortunate enough to be there and share in the team's delight as they discovered the ancient scuffled footprints of hoofed beasts stamped into the sand.

Tom's recent fieldwork at Gwithian and his analysis on land molluscs, alongside studies on pollen and buried soils by his colleagues, is presented here in full for the first time. This new work is much welcomed and is a major advance upon existing knowledge by providing us with forensic and detailed insights into the key palaeoenvironmental events and stories related to the excavated Gwithian "sites". The techniques and data collation methodologies that Tom and his colleagues have drawn upon had not been routine field practice 50 years ago (although Geoffrey Lewis did collect early samples of land snails from Gwithian in the 1960s). But these techniques were not routinely available to Charles and his team even if the research value of ancient molluscs was recognised for palaeoenvironmental archaeology was still in its infancy. Tom's research has however made good use of the chronological and sites' narratives established by Charles' earlier work and the constructive insights gained is a testimony to the robust framework established by Charles. Tom's research has been targeted and fruitful and gives us a more nuanced insight into the human story of people and land, farming and settlement at this magical coastal place from early prehistory to historical times. The new OSL and radiocarbon dates give us a tighter time frame and we learn that the impact of people at Gwithian from early prehistory is marked. The wet marshy open landscape of the Red River valley in earlier prehistory changed quite dramatically by the earlier Bronze Age when human settlement made its first firm foothold and sand blows began to make an impact on human settlement in the area. The late Bronze Age settlement in all its variety developed in a more stable landscape although its impact was probably more intensive, extensive and even pragmatic given thin agricultural soils and a background hazard of sand blows – which eventually may have forced or precipitated final abandonment. The recent discoveries of Bronze Age walls eroding at Strap Rocks shows how Gwithian will no doubt continue to yield future archaeological treasures and secrets as well as highlighting its ongoing vulnerability as a landscape. Tom's research, in the spirit of all good research, has enlarged our understanding, confirming some former working ideas and challenges others. Whether or not the Red River was always estuarine rather than intertidal is interesting in the light of the challenge of understanding the exact character of the enigmatic stone huts which form the post Roman settlement built in the dunes on a spit of land at Gwithian.

With its exotic post Roman imported pottery, Merovingian belt buckle and grass-marked pottery, this enigmatic "settlement" of small circular buildings has continuing relevance particularly with regard to current research on the post Roman citadel at Tintagel, which, with its distant maritime trading networks, like Gwithian, would have been linked into the wider world of Dumnonia. However stable Gwithian may appear today the landscape here is under constant pressure from change and it is a fragile resource and precious archaeological asset. Tom's new research just re-emphasises the regional and national importance of these special places and how safeguarding these coastal landscapes remains our present and future challenges.

Acknowledgements

This work is based on a PhD thesis undertaken in the Department of Archaeology, University of Reading, and I wish to thank the many people who have helped and supported me; if anyone is inadvertently omitted then I ffer my apologies.

Special thanks go to the late Professor Charles Thomas, the owner of much of the land at Gwithian, and to whom this work is dedicated. He readily granted access to his property and took great interest in the study.

Other land owners, Cornwall County Council and Mr & Mrs A. James at Gwithian, are also thanked for permission to work on their properties. Appropriate consents for fieldwork were obtained from English Heritage (Scheduled Monument CO 771) and Natural England (St Agnes to Godrevy Head SSSI), to whom thanks are given. The work has been continuously supported by Jacky Nowakowski of the Cornwall Archaeology Unit.

My main PhD supervisor, Professor Martin Bell, has actively encouraged my interests in molluscs; he initially suggested the project undertaken at Gwithian, and has supported me throughout with advice and friendship. He has provided many helpful comments on early drafts of this manuscript. My second supervisor, Dr Stuart Black, has been invaluable with advice on the appropriate laboratory methods and their interpretation.

Many staff and students in the University of Reading have been involved in both the field work at Gwithian and the laboratory work:

Field work: Prof. Martin Bell, Kerrie Bull, Dr Christine Bunting, Dr David Dunkin, Dr Jennifer Foster, Nick Harper, Richard Kevill, Simon Musset, Tom Rose-Jones, Dr Chris Speed, Dr Victoria Taylor, David Thornley, Kevin Williams.

Also Chris Verran, Cornwall Archaeological Society.

And Steve Hebdige and Precilla Oates for identifications of flowering plants.

Laboratory work: Yan Gao, Karen Gutteridge, Franz Street, David Thornley, Kevin Williams.

Grateful thanks also go:

Professor Helen Roberts, Aberystwyth University: Optically stimulated luminescence dating.
Dr Christine Davies, Aberystwyth University: Itrax XRF core scanning.
Dr Rowena Banerjea, University of Reading: Micromorphology.
Dr Rob Batchelor, University of Reading: Pollen analysis and report.
Dr Nigel Cameron, University College London: Diatom analyses.
Dr Jennifer Foster for the graphic of the worked flints in the Gwithian Trench.
The National Environment Research Council for the grant enabling radiocarbon dating of samples from Gwithian.

Finally, I thank my wife Sue for great support and encouragement throughout the project, and for accepting the disruption caused by numerous sand samples in our house.

Note concerning dates used in this work

Radiocarbon and optically stimulated luminescence dating was obtained on samples. The former are expressed as 'cal BC' or 'cal AD' whereas the latter are absolute dates that do not require calibration. In the discussions concerning chronology there is therefore a mixture of calibrated and absolute dates. Where quoted dates refer solely to one or the other dating technique the appropriate terminology is used, but when discussion involves dates obtained by both methods then 'cal' is omitted.

Photographs

Acknowledgment for photographs is given when necessary. All unacknowledged photographs are by the author.

Chapter I – Introduction

> The coastal sand dune sequences that line the exposed, gale-prone Atlantic shorelines of north-west Europe are the result of poorly-known histories of storminess and quiescence, geomorphic and pedological processes, and past human activities.
>
> (Gilbertson *et al.* 1999)

Coastal sand dunes are common around the coasts of Britain and northern Europe and are likely to have been visited regularly by humans during the Holocene. Sandy bays permit access routes for sea transport and to food resources both on land and from in-shore waters. Many archaeological sites may have been lost over time as a result of coastal erosion or sea level rise, while others were buried by blown sand and remain remarkably well preserved.

Dune formation is thought to have commenced during the Devension Glacial, although many dunes are below modern sea level. In some areas of Britain Pleistocene dunes exist onshore as raised beaches, as at Brean Down in Somerset (ApSimon 2000; Currant *et al.* 2006), Saunton and Croyde in Devon (Campbell and Gilbert 1998) and Newquay and Godrevy in Cornwall (James 1994, 1995).

These glacial dunes are generally only exposed in cliff sections and are not apparent on the land surface. The majority of Holocene dunes around the British coast did not appear until sea levels became relatively stable at slightly below present levels during the early Neolithic (Carter 1988, 303; Shennan and Horton 2002, 518).

ARCHAEOLOGY AND COASTAL DUNE SITES

The archaeological sites that survive in or under coastal dunes are an important resource for understanding human presence and activity in prehistoric periods during the Holocene. The blown sand which buries the sites may accumulate slowly or rapidly; stratigraphy is well preserved and less likely to be disturbed by later ploughing or other human activity than on sites further inland. Repeated episodes of sand blow may cover sequential phases of activity or occupation, the latter representing phases of stability. Bell and Brown (2008) have outlined the key contexts in which archaeological and palaeoenvironmental evidence is preserved on dunes (Figure 1). Human presence may be represented by the remains of buildings, field walls and trackways, for example, while their activities may be seen by preserved plough marks or animal footprints, charcoal from burning, midden material and artefacts.

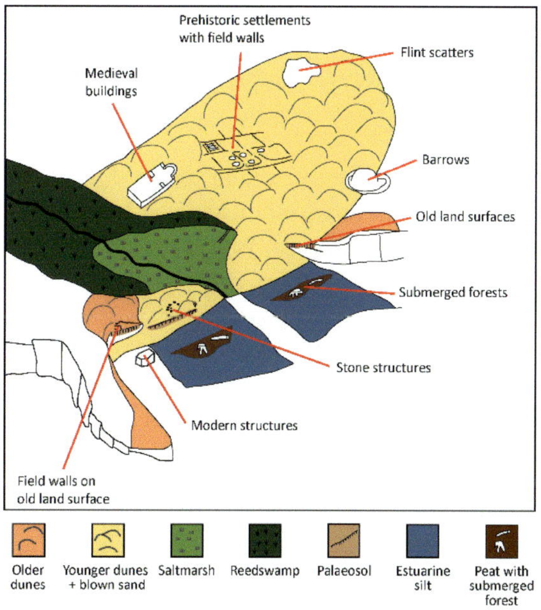

Figure 1. Diagram to illustrate some of the key contexts in which archaeological and palaeoenvironmental evidence is found in relation to coastal dunes. (redrawn from Bell and Brown 2008, 22)

Major sand blow episodes in different parts of Britain are likely to be broadly coeval, especially at times of rapid climate change, but less severe episodes will more probably impact only in local areas. Comparison of blown sand sites can aid in establishing chronologies of coastal environmental change, but only when there has been a robust programme of context dating in association with material culture and/or palaeoenvironmental evidence.

One of the most important sites where these sequences have been studied is at Gwithian in Cornwall, where excavations from the late 1940s to the mid-1960s showed three distinct phases of occupation during the Bronze Age (*c* 1800 cal BC, *c.* 1500–1200 cal BC and *c.* 1300–900 cal BC) followed by abandonment for around 1300 years before establishment of a post-Roman settlement on a nearby dune in the fifth century AD (Thomas 1958; Megaw 1976; Nowakowski *et al.* 2007). There was a Roman homestead on the adjacent headland (Fowler 1962) but no known settlement on the dunes at Gwithian during the Iron Age or Roman period. On the neighbouring hill – Godrevy Towans – a mid to late medieval manor was established overlying an Iron Age round. Little palaeoenvironmental analysis was undertaken during these excavations, and a trench was opened in 2005 within the Bronze Age settlement for this purpose, although the findings have not been fully reported (Nowakowski *et al.* 2006). The present study aims to address some of these palaeoenvironmental issues, mainly by the study of molluscs, and attempts to determine how episodes of sand blow may have contributed to the episodic settlement and abandonment of the Gwithian area. These events can be related to known climatic events during the Holocene: warm periods (Johnsen *et al.* 2001; Mann *et al.* 2009; Wang *et al.* 2012); cool periods (Barber *et al.* 1994; Van Geel *et al.* 1996; Barber *et al.* 2000; Fagan 2000; Barber *et al.* 2003; Mann *et al.* 2009; PAGES 2k Consortium 2013); instability (Bond *et al.* 1997; Bond *et al.* 2001) and flooding events (Johnstone *et al.* 2006; Macklin *et al.* 2006).

Cornwall was a major source of tin during early historic times, becoming a major industry in the eighteenth and nineteenth century. The Red River, adjacent to the archaeological site, flows from the Camborne and Redruth mining areas, and carried much of the mine effluent, leading to the river's current name. Deep coring in the valley of the river allows assessment of possible mining in prehistoric times, and also to establish when the river received its modern name, changing from the earlier Conor Dour.

MOLLUSCS AND ARCHAEOLOGY

Terrestrial molluscs living in Britain are generally small in size, the majority being less then 10mm in maximum length/width, and many less than 5mm. Their small size and rate of movement means that they are restricted to relatively small areas of ground during life, and that the site where they die is likely to be a good indicator of the habitat in which they were living. This does not necessarily hold for an assemblage found some time after death, as the dead shells may have moved large distances, especially as a result of wind and/or water action. When interpreting any death assemblage, it is important to consider whether that assemblage is autochthonous, *in situ*, and representative of the habitat during life or allochthonous, not *in situ*, and displaced from the site inhabited while alive.

Despite these restrictions, knowledge of the varying ecological preferences can provide useful information concerning the local habitat at the time when those molluscs were living. The varying ecological preferences of different taxa are well recognised (e.g. Boycott 1934, 1936; Sparks 1959–60; Evans 1972), being divided into 'shade preferring', 'catholic' or 'intermediate' and 'open country' for terrestrial shells and 'slum', 'ditch', 'moving water', 'catholic' and 'brackish water' for aquatic species. However, it is important to stress there are serious limitations to this rather simplistic approach (e.g. Cameron 1978; Thomas 1985). This autecological approach takes the preferences of individual species using the broad groupings given above to recreate the habitats and takes little account of interactions between species. It assumes that modern preferences are no different from those in earlier periods, the principle of uniformitarianism. For some species this assumption is known not to hold, for instance that of *Lauria cylindracea*, where Evans (2004) demonstrated that in prehistoric times this species favoured shady, or least diverse, habitats, but in the last 2000 years has become associated with open grassland and sand dune habitats.

A synecological approach takes a broader view of environmental reconstruction from mollusc analyses, in that it considers interactions which cross the traditional habitat boundaries. The concept of taxocenes has been developed, which recognises associations of species found in living communities, and develops the ecological groupings of species which may be applicable to sub-fossil assemblages. Such taxocenes have been proposed for both dry and wet ground (e.g. Ložek 1990; Evans 1991; Evans *et al.* 1992; Davies *et al.* 1996; Davies 2008).

Non-marine molluscs have been increasingly used in palaeoenvironmental analysis of archaeological sites over the last century (see Evans 1972; Davies 2008). These have concentrated mainly on the inland chalkland landscape of Britain (see Allen *et al.* 2009, and references therein), with relatively few examining dune sites (Figure 2). The latter are largely restricted to the south-west of England, Pembrokeshire in Wales, and the Scottish Islands (the Outer Hebrides and the Orkneys). Details for each individual site are given in Walker (2017), and only a few key sites are briefly discussed here.

Towan Head, Newquay, Cornwall

Mollusc analyses in the first decade of the twentieth century have been reported from buried soils with overlying blown sands within a Holocene sand cliff at Towan Head (Kennard and Warren 1903; Woodward 1908) and are important as they are among the earliest studies to use shells in palaeoenvironmental reconstruction, although they were not associated with any archaeology in the immediate area. These deposits have been revisited in later studies (Spencer 1975; Milles 1991; Walker 2014), but no radiocarbon dates have been obtained.

The basal layers show a shaded environment with a range of species indicating light wood or scrub. Although absent in the lowest deposits, *Pomatias elegans* was found above the basal old ground surface; this is a species of highly calcareous soils, associated with shady, moist habitats, and now known living in only a single area of Cornwall, just south of this site (Turk *et al.* 2001, 6). Its presence suggests some form of disturbance such as woodland clearance (Evans 1972, 133), which in this context may have been during the Neolithic and associated with forest clearances. The wood/scrub gave way to open country as blown sand accumulated, with the appearance of typical xerophile species such as *Cernuella virgata*, *Helicella itala* and *Cochlicella acuta*.

Brean Down, Somerset

The Bronze Age settlement at Brean Down is arguably the most closely dated dune sequence in Britain, being excavated in the 1980s (Bell 1990). There was evidence of Mesolithic and early Neolithic presence (flints) prior to the settlement, and continuing activity into historic times, with early medieval graves and a post-medieval building. Chronology was established with a sequence of twenty one radiocarbon dates from bone, peat and charcoal, which were correlated with five mollusc columns forming a continuous stratigraphic sequence from the settlement and a test pit (Spencer 1974; Vaughan 1976; Bell and Johnson 1990; Walker 2008).

The basal palaeosol was decalcified, but other environmental evidence, particularly micromorphology, indicated that early Holocene woodland was cleared by burning during the Neolithic; the land was cultivated before being buried by blown sand in the mid fourth millennium BC. The molluscs in a Beaker period stabilisation horizon showed open conditions (*Vallonia* sp., *Helicella itala*, *Truncatellina cylindrica*), but the presence of damper, shadier, habitats was indicated by clausiliids (*Clausilia bidentata*, *Macrogastra rolphii*, *Balea perversa*) as well as *Carychium tridentatum*, *Acanthinula aculeata* and *Aegopinella nitidula*.

Figure 2. Coastal sand dunes in the British Isles, shown in red.
The sites which include molluscs in palaeoenvironmental analyses are named. (modified from Doody 2008, 2009).

Following a period of unstable blown sand, stability returned during the middle Bronze Age, associated with the main period of settlement, but with the introduction of *Cochlicella acuta* and *Cernuella virgata*, although the question of intrusion of material from more superficial levels is raised, as *Cernuella* is generally considered to be a Romano-British introduction (Kerney 1999, 181; Davies 2010, 176). The presence of brackish water species during the middle Bronze Age indicated the presence of estuarine mud flats. After the Bronze Age the landscape became progressively more open, up to the sixteenth/seventeenth centuries AD, with a diverse assemblage showing greater vegetation cover being found in twentieth century levels.

Northton, Isle of Harris, Outer Hebrides

An early Mesolithic to post-medieval settlement site was excavated in the mid-1960s and in 2001 (Simpson 1966; Murphy and Simpson 2003; Murphy *et al.* 2004). There were multiple well-stratified occupation horizons interspersed with blown sand all within about five metres of windblown calcareous sand. Multiple radiocarbon dates establish the period of occupation, from the Mesolithic (Gregory *et al.* 2005) to the post-Beaker and later Iron Age (Murphy *et al.* 2004). Mollusc sequences were analysed by Evans (1971b; 1972, 294; Law and Thew 2017, 87), although interpretation was complicated by the effects of deflation and rapid deposition of sand by wind action, with subsequent loss of shells and incorporation of molluscs from a variety of habitats. Despite these limitations, Evans established a comprehensive pattern of habitat change over time.

Molluscs were absent from the Mesolithic and early Neolithic layers, but during the mid-Neolithic they indicated a woodland environment with *Cochlicopa* spp. and *Vitrina pellucida* dominating, but with a range of other hygrophile and shade-loving species. The woodland was then cleared, indicated by a marked rise in open country species (*Vallonia costata, Vertigo pygmaea, Pupilla muscorum*). Following a brief period of open country there was woodland regeneration during the later Beaker period, stability being shown by a marked increase in numbers of shells, and a fall in the proportion of open country species. The second woodland phase ended with clearance, with sand accumulation increasing during the Iron Age; the molluscan fauna now comprised largely the open country species mentioned above, together with *Vallonia excentrica*. Only towards the end of the Iron Age was stability established, with a marked shift in mollusc taxa, becoming almost entirely limited to *Cochlicella acuta* and *Helicella itala*, both species typical of open dune grassland. Regarding the periods of deforestation, Evans (1971b, 59) postulates that 'In both cases, clearance was probably for pasture as none of the sherds from the occupation horizons bore grain impressions nor were there found querns, sickles or other trappings of an arable farming community.' However, he does not rule out (1972, 296) that the destruction of the woodland was not 'an altogether natural process brought about by the overwhelming action of the sand.'

The Bay of Skaill, Mainland, Orkney

Storms in the Orkney Islands in 1850 revealed stone buildings eroding out of the cliffs on the west coast of Mainland, revealing the Neolithic village of Skara Brae in the Bay of Skaill. It was not until the 1920s that full excavation was carried out under the direction of Gordon Childe (1931). The buildings provide an unrivalled example of Neolithic settlement, and are a wonderful example of how archaeological sites can be preserved under blown sand deposits.

Although the original excavations predated modern concepts of palaeoenvironmental analysis, mollusc studies have since been performed at Skara Brae (Spencer 1974, 1975; de la Vega-Leinert *et al.* 2000). There was clear evidence of original open woodland or scrub (*Cochlicopa* spp., *Lauria cylindracea, Clausilia bidentata, Oxychilus alliarius, Aegopinella nitidula, Vitrina pellucida*), but with a strong open country element (*Vertigo pygmaea, Pupilla muscorum, Vallonia excentrica*); this woodland/scrub then gave way to purely open country and catholic species, possibly a result of anthropogenic clearance. On the north side of the Bay of Skaill bore hole studies indicated that there was initial ponding in dune slacks during the woodland phase, shown by the presence of several freshwater mollusan species (de la Vega-Leinert *et al.* 2000). This was followed in the later Neolithic with multiple episodes of sand blow on open grassland, interspersed with ponding and a rising water table.

THE PRESENT STUDY

The majority of archaeological studies that incorporate palaeoenvironmental analyses, and in particular molluscs, tend to concentrate on the areas of primary concern, usually in or very close to the sites being excavated. Relatively few explore the wider landscape or attempt to place settlement activity in the context of environmental change beyond the bounds of the area of immediate interest. The study described in this monograph explores a wide area around the main occupation sites, showing evidence of human activity over a span of around 8000 years, in a landscape that is now blanketed by sand. A wide variety of environmental and geoarchaeological modalities have been employed to examine change over time, in an attempt to establish how changes in the environment have influenced human presence and activity, and how, in turn, that activity may have altered the local landscape.

Molluscs have been used in this study as the main palaeoenvironmental indicator, obtained both from mollusc columns and from core samples. There are particular problems in mollusc analyses in dune environments resulting from the instability of sand. Winnowing during windy conditions can rapidly cause deflation of assemblages, either with loss of entire stratigraphic sequences or with removal of lighter shells or fragments leaving a biased residue of heaver material. Equally, conflation of small or large deposits of sand in stormy weather may result in allochthonous assemblages that are not representative of the area under consideration. These problems are discussed in the relevant chapters.

EXTINCTION AND INTRODUCTION OF MOLLUSC SPECIES

The fauna of Britain is not constant. Some species become extinct either through natural causes or the action of people, while other are introduced, often inadvertently but sometimes deliberately. The aurochs, lynx and brown bear are examples of mammals that no longer inhabit Britain, while fallow deer, rabbits and rats are introductions during the Holocene (Kerney 1966; O'Connor and Sykes 2010). Molluscs are no exception. Many species survived from the last Ice Age and are still found in Britain, although some with very restricted distributions. A few species did not survive the warming during the early Holocene and are now extinct in Britain. Others have been introduced (Davies 2010; Cameron 2016, 371). Introductions may be accidental, arriving, for example, on garden produce, although a few species were almost certainly brought to Britain intentionally; an example is the Roman snail, *Helix pomatia*, which is thought to have been a food resource during Roman times (Davies 2010).

If the timing of extinctions and introductions can be established with a high degree of certainty, then this can be useful from the chronological point of view in archaeological deposits. It is, however, very important to consider how secure is the stratification, as the presence of molluscs in unexpected strata is frequently due to intrusion, often by worms or rodent activity. Dating correlation should, whenever possible, be obtained by independent testing rather than relying on the presence or absence of molluscs, although the latter may be helpful in some circumstances. For example, if *Xerocrassa geyeri*, a species now extinct in Britain, is present in a sediment then that sediment almost certainly dates to the very early post-Glacial period (Kerney 1999, 184); only at Gwithian in Cornwall has it been found in horizons dating to the Bronze Age.

The introduction of particular species is perhaps more useful from the chronological point of view, although the problems of bioturbation need to be taken into consideration, and shells found in inappropriately deep levels may be ascribed to older sediments than they should be. For example, the various species introduced into Britain in medieval or later periods should not be present in deposits considered to date from earlier centuries or millennia, and if present at inappropriate early dates probably indicate bioturbation or other forms of contamination. However, the *caveat* is that not all is yet known concerning introductions and extinctions of many species, and detailed work may change present understanding of the time scales, as the present study shows.

Chapter II – Gwithian and its archaeological complex

The modern village of Gwithian lies on the north coast of Cornwall on the east side of St Ives Bay, 5km north-east of Hayle, and 7km north-west of Camborne (Figure 3), which was, until recently, the centre of the densest area of tin mining in Britain. A settlement in the area was recorded in 1086 in the Domesday Book (Figure 4), when it was called 'Conarditone'.

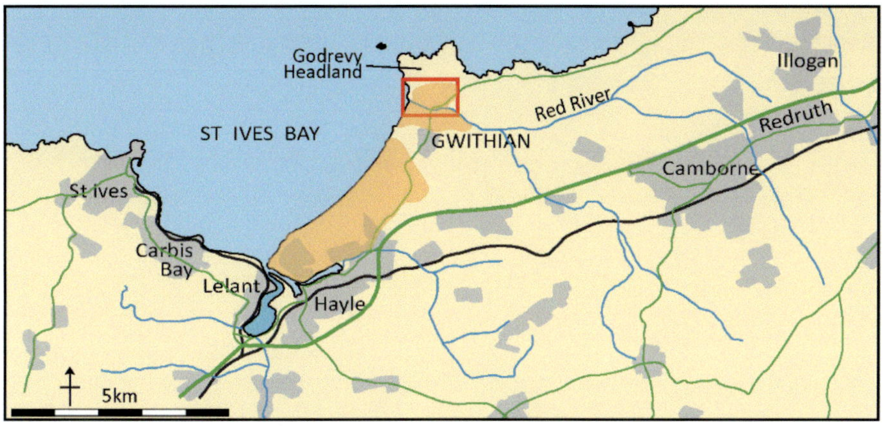

Figure 3. The location of Gwithian and the study site (red box) in west Cornwall. Sand dunes between Hayle and Gwithian are highlighted in orange.

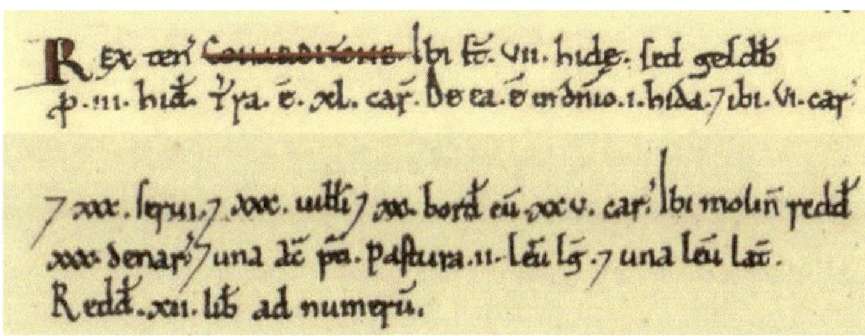

The King holds Conarditone. There are 7 hides, but it paid geld for 3 hides. There is land for 40 ploughs. Of this, 1 hide is held in demesne, and there are 6 ploughs and 30 slaves: and 30 villains and 20 bordars with 25 ploughs. There is a mill rendering 30 denarii, and 1 acre of meadow, pasture 2 leagues long and 1 league broad. It renders 12 pounds by tale.

Figure 4. Entry for Gwithian in The Domesday Book.
(original: http://opendomesday.org; accessed 13 June 2017)
(translation: Williams and Martin 2003, 342)

In medieval times it had various names, Conerton, Conarditon or Nikenor, all with spelling variations, although these names may refer to an older settlement a short distance away from the modern village (Thomas 1964a). In his perambulations around Cornwall in about 1538, John Leland mentions that the settlement was 'now gone' – presumably buried in sand:

> Nikenor a 2. miles from Ryvier sumtyme a great Toun now gone. 2 Paroche Chirchis yet seene a good deale several on from the other, sumtyme in the Toune, but it is now communely taken to be in S. Guivian's Paroch, and there commith a Broket to the Sea.
>
> (Hearne 1769, vol. 3: 18)

There is now no trace of this 'great town' except for the remains of St. Gothian's chapel on a slight rise of ground about 700m north of the present church (Thomas 1964a).

To the north of the village the lower valley of the Red River contains dunes and areas of blown sand close to its outflow into the bay. The river arises in the hills around Camborne, and is about 13km in length. Extensive sand dunes line the coast from Hayle to the Red River valley and were exploited for sand extraction for many years (Nowakowski *et al.* 2007, 15). North of the river is Godrevy Headland, with precipitous cliffs up to 65m in height. Godrevy Towans is an area of high ground in the southern part of the headland (Figure 5) and is covered by a surface layer of blown sand. The term 'Towans' derives from the Cornish *towyn*, a 'turfy down' (Edmonds 1846, 301), perhaps simply with change of the 'd' of 'down' to the 't' of 'towyn'.

Figure 5. Aerial view of the Godrevy Headland, looking north-east.
The Towans are the area of high ground in the centre of the photograph.
The Bronze Age and post-Roman settlements are at the east end of the meadow in the lea of the Towans.

GEOLOGY AND GEOGRAPHY

The underlying solid geology (Figure 6) is Devonian, with the Porthtowan Formation derived from the Gramscatho Group of interbedded sandstone, slate and siltstone and with areas of exposed Mylor Slates (British Geological Survey 1984; Goode and Taylor 1988; Shail 1989), known locally as killas, an old Cornish miners' term. Quaternary head lies over some of the slates and this is topped by the blown sand of the Towans. The sand in the Red River valley extends to a depth of about 10m and there are alluvial deposits along the line of the river.

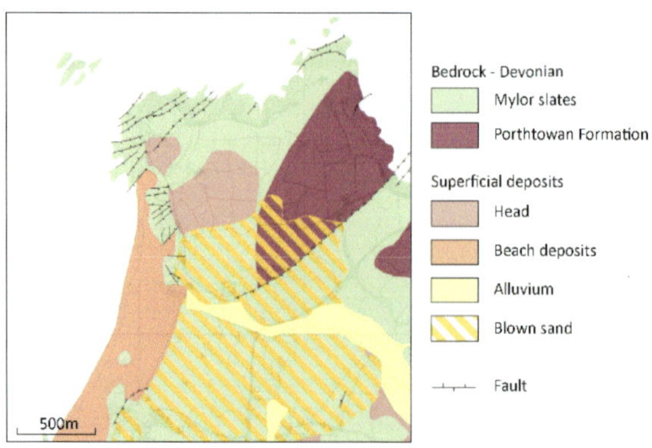

Figure 6. The solid and superficial geology of the Gwithian/Godrevy area.

Sections of the cliff at Godrevy (the exact points were not specified) have been reported by Robson (1943–44, 147) and Thomas (1957–58, 7) and show the presence of a raised beach on the slates with head and blown sand above it. This Pleistocene raised beach is visible in many cliff sections both south and north of Godrevy beach (Figure 7), consisting of well-rounded pebbles up to about 20cm in diameter, 4–5m above modern sea level. In places this beach may be about 1m in thickness. Above these raised beaches along a 100m length of Godrevy Headland are outcrops of calcite-cemented shell sand known as sandrock (Campbell *et al.* 1999) (Figure 8), found on the north Cornish coast only at Godrevy, Fistral (Newquay) and Trebetherick (Howie and Ealey 2010). This hard compacted stone has been used for many centuries as building material (e.g. Borlase 1758, 95).

Figure 7. The raised beach at Strap Rocks.

Figure 8. Sandrock boulders fallen from the cliff on Godrevy Headland.

The Red River runs to the sea in a flat valley with Gwithian village to the south and Godrevy Towans to the north. This river, approximately 13km in length, arises near Bolenowe west of the granite mass of Carnmenellis and, with several tributaries, drains into the east side of St Ives Bay (Figure 9).

Figure 9. The lower Red River valley viewed from near the summit of Godrevy Towans, looking south. St. Ives Bay is on the right, with Gwithian/Phillack Towans between the sea and the road. The Red River runs from left to right across the centre of the photograph.

For much of its inland course the river runs through the Camborne/Redruth tin mining area and it is likely that its current name originates from the colour of its water which became stained a rich red from the ochre released during tin extraction (Figure 10). The original name of this river was Conar Dour (with many spelling variations), meaning 'river of fury'. Borlase (1758, 262) states 'in the river Conar, which divides the parish of Camborne from Gwinear and Gwythien ... the many mines which have been of late years wrought in the neighbourhood, have destroyed the fish.' By the 1790s it had become known as the Red River (Hamilton Jenkin 1965, 24), so the river may have been renamed between these dates. After cessation of mining in the Camborne area in the 1990s the river was restored to its natural clean state.

Figure 10. The outflow of the Red River into St Ives Bay during the active mining era.
(© Historic England Archive; ref. AFL03/Aerofilms/363441)

The lower reaches of the river were considerably modified in the nineteenth century. In 1806 the outlet to the sea was at the south end of Godrevy beach (Figure 11) and was unchanged on the Ordnance Survey First Series map of 1830. By 1839 (Figure 12) its outflow had been diverted adjacent to Godrevy Towans at the north end of the beach and by 1850 the length immediately upstream and downstream from Gwithian Bridge altered at least twice, evidenced by 'New cut' and 'Former cut' on an 1850–51 estate map (Figure 13). A drain or leet was cut north and parallel to the river from just below Gwithian Bridge upstream for 1.5km sometime between 1839 and 1850. The reason for its creation is not recorded although it may serve to reduce flooding (Kirkham 2005, 14). Another possibility is that it was capable of providing clean, clear, water for human and animal consumption, as opposed to the mine-polluted water of the main Red River.

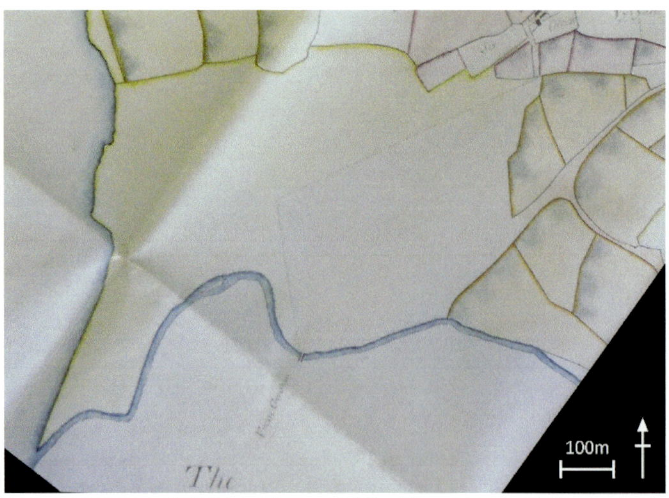

Figure 11. Part of Tehidy estate map of 1806 by James Mills. The course of the Red River is shown prior to its diversion. (Cornwall Record Office; ref. AD894/7/8)

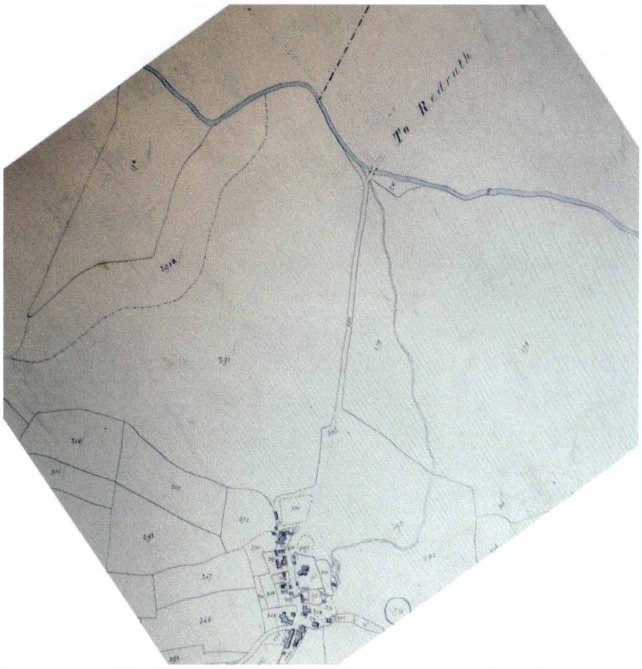

Figure 12. Part of Tehidy tithe map, 1839. The Red River now opens more directly into St Ives Bay on the extreme left of the image although the original river course is still shown. (Cornwall Record Office; ref. P81/27/2)

The 1850–51 estate map (Figure 13) marks 'Sand Pits' north of Gwithian Bridge, evidence that sand extraction has taken place in the past, which may have altered the modern landscape more than is easily recognised. Fowler and Thomas (1962) state that sand extraction was practiced perhaps as early as the seventeenth century, the sand being used for liming the fields in the area, this practice also being documented by Roberts (1910–11). The Spring Assizes of 1834 recorded that farmers had been using towan sand for manure, and had been doing so from 'times beyond memory of man' (Rowe 1953, 219).

There is good place name evidence that the lower Red River valley has been marshland for up to 3km from the sea, at least back until the late Middle Ages. The Cornish word for marsh is 'Hal' or 'Hel' and the place names Halglasen, Hellow, Halenvelyn and Halvernocks are recorded back to the mid-fifteenth century (Nowakowski and Thomas 2007, 56). It has been suggested that this was estuarine marsh rather than a purely freshwater valley (Stephens 1899, 1899–1900; Thomas 1958; Nowakowski and Thomas 2007, 57).

Figure 13. Part of 1850–51 Tehidy estate map by Nicholas Whitley.
The 'New cut', 'Former cut' and 'Ancient course of the river' are shown. (Charles Thomas collection)

The deposits along the valley were investigated at the end of the nineteenth century when about 200 bore holes were put down along and across the valley to assess their potential for mineral rich deposits (Stephens 1899, 1899–1900). Unfortunately, detailed reports of these cores do not seem to have survived, although some drawings are extant (Figure 14). F. J. Stephens, who conducted this coring exercise, concluded that 'the sum of the borings went to prove the former existence of a shallow tidal estuary with no deep channel or old river bed in the centre as was expected' (Stephens 1899). He based this assumption on the finding of stiff clay or consolidated mud deposits without any shells, bluish-grey in colour, which he concluded were 'evidently estuarine, and mark at least two periods in what was then a creek' which extended at least 1½ miles (2.4km) up the valley. Stephens' extant drawings show a line of six boreholes extending over a distance of 2954ft (900m) (Figure 15); the nearest to the sea being 12m east of Gwithian Bridge, where he found a thick band of clay at 1.7m below the surface, with bedrock at a depth of 6.1m. However, Hosking and Ong (1963–64) reported a borehole close to the road bridge (stated 'seaward' in the text, i.e. west of the bridge, but 'just east of the main coast road' in their table) and near the present course of the river; they only found a thin (0.3m) band of clay at 8.5m depth, with the killas at 10.7m. The difference between the two findings is not explained.

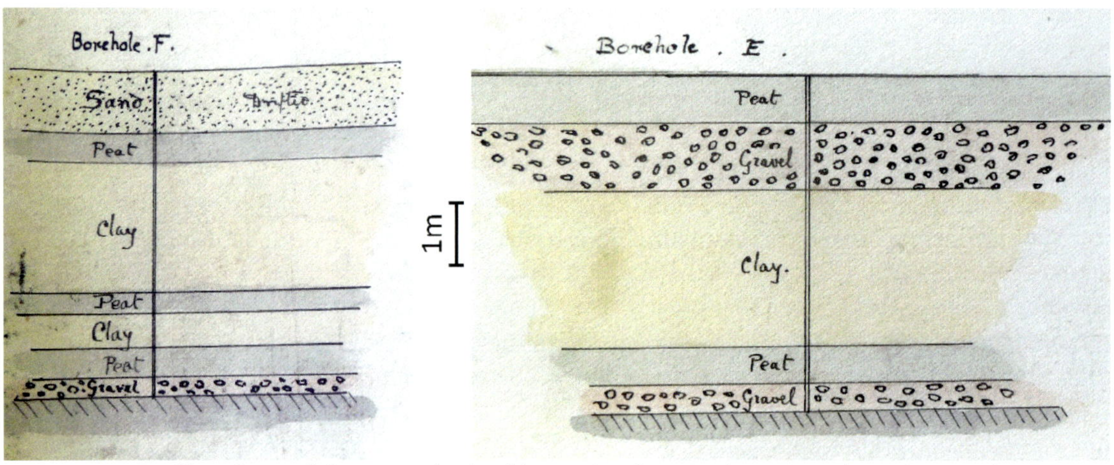

Figure 14. Profiles of two of the cores obtained by F.J. Stephens in the region of Gwithian Bridge.
Borehole F is 12m east of Gwithian Bridge and borehole E 235m east.
The transect line of the present study is between these two cores (180m east of the bridge).
The depth from surface to base of the gravel is 6.1m. (Cornwall Record Office; ref. MRO/A48)

During the 1980s commercial boring for the Central Electricity Generating Board and for a regional sewerage scheme was undertaken at several locations between the Red River and Gwithian village, (British Geological Survey 2013). Clay was found in most of these bores, at depths ranging from 1.2m to 8.4m (Figure 15) and at 7.35cm and 8.40cm in the region of the cores in the present study. In 2002 a survey of 30 auger holes was undertaken in the area south of the river, between the road and the beach, prior to the establishment of a nature reserve in an area of old sand pits (Wessex Archaeology 2002). Most of the cores only reached 3m in depth, showing thick layers of windblown sand; the deepest, at the site of the original river course, reached 5m and was the only one to show estuarine clay at 4.45m. In only two cores were there any silty layers consistent with old ground surfaces, at 3.78m and 3.80m respectively.

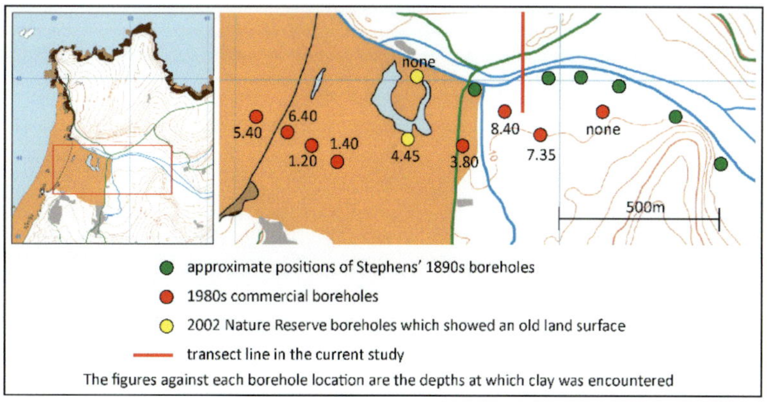

Figure 15. The locations of the boreholes in the lower Red River valley.

Windblown sand covers large areas of ground along the course of the lower Red River and the adjacent countryside (Figure 6), extending inland for at least for 1km, beyond which the valley becomes marshy. To the north sand covers Godrevy Towans, with only a thin layer of sand at higher elevations, but in some areas with dunes up to about 8–10m in depth. Between the river and the estuary of the Hayle River to the south the contiguous Gwithian, Upton, Mexico and Phillack Towans form an extensive dune system covering 294ha (Saye 2003, 550). The dunes are at present largely vegetated, with few areas of bare sand. Much of the lower Red River valley is used for grazing, mainly cattle but also sheep, and there is a flourishing rabbit and fox population.

ARCHAEOLOGY

The area around Gwithian and Godrevy is remarkably rich in archaeology, ranging from the Mesolithic to the medieval period, concentrated on the flatter land either side of the Red River, but little was recognised prior the middle of the 20th century. The medieval settlement of Conerton lay between the river and the present village of Gwithian, but is now buried by sand (Thomas 1964a). Some remains were still visible when Leyland visited around 1738, but when William Borlase arrived in the mid-18th century nothing remained visible apart from a few 'cotts' and 'dead bodies':

> It was formerly a considerable town … had two churches, reach'd to the river; this town is now buried in the Sand, all but a few cotts near the Southernmost church but as the sands are not high from the banks of the River to the present church, either the Land (where the Northern church [St Gothian's Chapel] appears by the dead bodies and other remains to have been) is sunk and the houses cover'd, or the blasts of sand render'd that part of the town uninhabitable, and the ruines must be removed into the hedges, and cotts, of the neighbourhood.
>
> (Borlase, quoted in Thomas 1964a, 6)

The archaeology began to be explored in detail by Charles Thomas, who led major excavation programmes both in the Red River valley and on Godrevy Towans and Headland (Figure 16) from the late 1940s into the 1960s. The locations of the main archaeological sites are shown on Figure 17. These excavations played an important part on establishing the archaeological potential of coastal blown sands and the identification of prehistoric field systems and traces of cultivation, as well as helping to develop a multi-phase and multi-site approach to landscape archaeology. Many of these ideas were later to become key themes in environmental archaeology. However, at the time, the resources and many of the techniques of environmental analysis were not available. The simple aim of this book it to fill that gap, and to complement the brilliant original fieldwork with a palaeoenvironmental story.

Figure 16. View looking north across the Red River valley with Godrevy Towans filling the background. The main Gwithian archaeological site is in the centre of the photograph at the base of the Towans

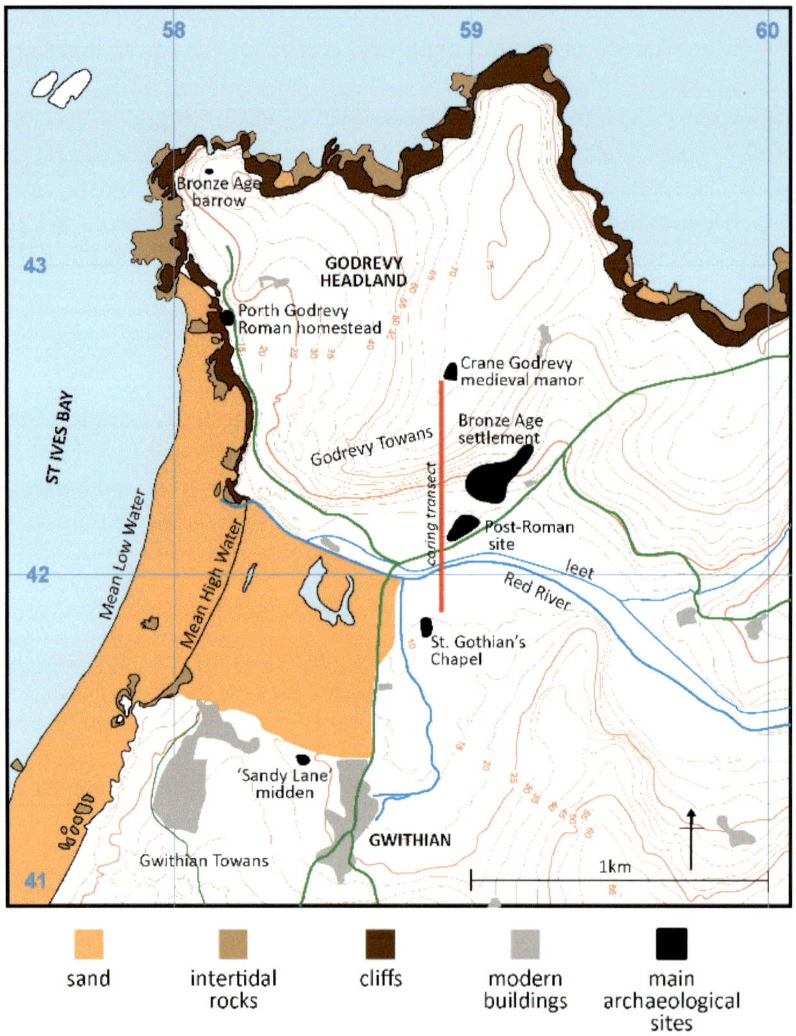

Figure 17. The main archaeological sites in the Gwithian area.
The coring transect line in this study is shown to indicate its relationship to the archaeological sites.

Thomas also collated numerous finds collected by field walking or shown to him by many other people. There were regular interim and summary reports of the excavations (e.g.Thomas 1958; Megaw *et al.* 1960–61; Fowler and Thomas 1962; Thomas 1969; Megaw 1976), and a summary in 2007 brings together much of the work, including 20 radiocarbon dates (Nowakowski *et al.* 2007), but the excavations have never been fully published. The entire Gwithian archive was examined and indexed between 2003 and 2007 following the award of a grant to the Cornwall Historic Environment Service by the Aggregates Levy Sustainability Fund (ALSF), administered by English Heritage. The results have been collated into a series of narratives covering the Bronze Age, post-Roman and later medieval excavations (Davies 2006; Sturgess and Lawson Jones 2006a, 2006c; Nowakowski 2007a, 2007b).

In 2005, as part of this post-excavation programme, a small trench (4.6m x 2.0m) was opened within the Bronze Age settlement to obtain palaeoenvironmental samples, and 20 samples for radiocarbon dating and two for OSL dating were obtained (Hamilton *et al.* 2008). The environmental samples were assessed (Nowakowski *et al.* 2006) but unfortunately full analyses could not be carried out because ALSF funding ceased to be available for archaeological projects.

Mesolithic

Many Mesolithic flints have been found in over 20 recorded sites in the Gwithian area (Thomas 1957–58; Roberts 1987; Nowakowski 2004, Figure 5; Thomas 2007, 21). No *in situ* flint is available in the vicinity but flint pebbles are found either in the Pleistocene raised beaches or in the shingle of the Holocene beaches.

Neolithic

There is very little evidence of Neolithic presence in the Gwithian area. Some early Neolithic gabbroic sherds, a few lithic pieces and two fragments of greenstone axes were found in early Bronze Age contexts indicating that there was some Neolithic activity in the vicinity (Megaw 1976; Quinnell 2007c, 23).

Bronze Age

A Bronze Age barrow mound on the north-west tip of Godrevy Headland was excavated by Charles Thomas and T. J. Scantlebury in the 1950s (J. Nowakowski, pers. com.) and re-excavated during 2012 (James Gossip, pers. com.). The findings of these excavations have not been published.

The major Bronze Age presence is a settlement site at the foot of Godrevy Towans towards the eastern end of the meadow north of the Red River (Figure 17). Excavations between 1954 and 1961 were led by Charles Thomas, whose principal assistants were Vincent Megaw, Bernard Wailes and Peter Fowler, all later to become leading archaeologists. These studies revealed three phases of occupation: *c.* 1800 cal BC, *c.* 1500–1200 cal BC and *c.* 1300–900 cal BC (Megaw *et al.* 1960–61; Sturgess 2007a; Nowakowski 2009). The following is a summary of the chronology, taken from Sturgess (2007a):

Phase 1	*c.* 1800 cal BC	**Early Bronze Age**. A homestead and farming; a single post-built structure within a fenced enclosure associated with terraces and ploughed fields (Figure 18).
Phase 2		Possible minor neglect: a blown sand horizon.
Phase 3	*c.* 1500–1200 cal BC	**Middle Bronze Age**. Settlement and farming; a probable stone and post-built structure associated with terraces and field boundaries and several phases of ploughed fields (Figure 19).
Phase 4		**Middle Bronze Age**. Farming and minor neglect of an isolated area; a blown sand horizon which contained midden material, suggesting soil improvement.
Phase 5	*c.* 1300–900 cal BC	**Middle to late Bronze Age**. A major settlement with farming (Figure 20), fishing and craft industries.
Subphase 5a		A series of roughly dug hollows, possibly the result of vegetation clearance.
Subphase 5b		A new farmstead. A group of post-built structures set within fields, with associated boundaries and terraces and phases of ploughing.
Subphase 5c		A group of stone-built enclosures and structures associated with field boundaries and ploughing.
Subphase 5d		Settlement demise. Phased dumps of midden material overlying and in the vicinity of the post-built and stone-built structures.
Phase 6		Sand dune formation: a multi-phase blown sand horizon containing sparse evidence for later activity.
Phase 7		Modern turf and topsoil.

Figure 18. View of a phase 1 early Bronze Age building (site GMXV).
Wooden stakes have been placed along the line of the enclosure and at the entrance of the structure.
(© J. V. S. Megaw, Gwithian Archive)

Figure 19. Bronze Age plough marks from phase 3 (site GMX).
(© J. V. S. Megaw, Gwithian Archive)

Figure 20. Close-up view of spade marks in the phase 5 Bronze Age sequence.
(© J. V. S. Megaw, Gwithian Archive)

One of the remarkable discoveries was the evidence of cultivation at Gwithian found during all phases of the Bronze Age occupation (Figure 21). Large areas containing well preserved plough or ard-marks were revealed in both the middle and late Bronze Age phases (Figure 19) (Thomas 1958; Megaw *et al.* 1960–61) and are among the first such evidence of cultivation to be described in Britain. It is of interest, in the light of the findings in the trench opened during the present study (see Chapter X), that no cattle or ovi-caprid hoof marks were found during excavations on the Gwithian Bronze Age site.

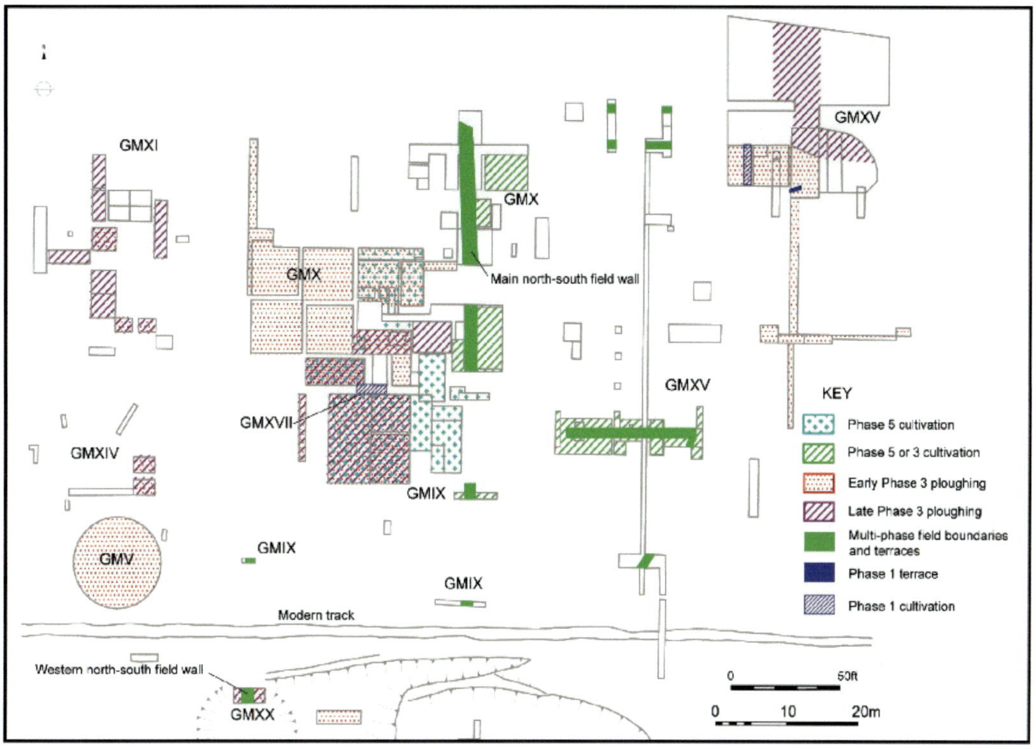

Figure 21. Schematic plan of the multi-phase Bronze Age fields and boundaries at Gwithian. (© J. Sturgess, Gwithian Archive)

There is good evidence for Bronze Age metal working at Gwithian. A socketed axe mould was found during the 1956–61 excavations within a building on site GMX (for location see Figure 21) which is considered to date from the 12th–11th centuries BC (Burgess 1976); this is among the first true socketed axes to be developed in the British Isles (Schmidt and Burgess 1981, 173). It is probable the metals necessary to produces axes were obtained from local sources in the Camborne area (see Chapter XIII).

An early Bronze Age gold lunula was found in Penwith in 1783 (Figure 22, left); its discovery site is unclear, but Trevarnon Round, 1km south of Gwithian, is a likely location (Mattingly *et al.* 2009). As recently as 2010 an unfinished gold torc from the middle Bronze Age was found 'on Gwithian meadow land', but the discovery site has not been made public (Figure 22, right).

Figure 22. The gold lunula (left) found in 1783 and the torc (right) in 2010. (lunula – © Trustees of the British Museum: object 1838.0519.1,AN25106001; torc – © Royal Cornwall Museum: object TRURI 2013.3)

Iron Age and Roman

Activity in the Gwithian area reduced considerably after the Bronze Age (Quinnell 2007b). An Iron Age/Romano-British univallate earthwork is present on the summit of Godrevy Towans, beneath the medieval structures of Crane Godrevy (see below). A Roman period homestead at Porth Godrevy, close to the cliff edge on the west side of the headland, was excavated in 1956–58 (Fowler 1962). This was a single dwelling within an enclosure; dating of the settlement is unclear, but there was probably occupation from the late first/early second century into the fourth century AD, before abandonment for unknown reasons. There is a Roman villa dating from the late second to fourth century AD about 5km to the east of Gwithian at Magor (O'Neill 1934). Scatters of gabbroic pottery from the Roman period have been found south of the Red River, but in such low frequency that deposition of waste on agricultural land has been suggested (Quinnell 2007b, 38).

Post-Roman

The major post-Roman activity in the Gwithian area is on an elevated dune slightly to the south-west of the Bronze Age site (Figure 17). This was excavated throughout the 1950s and showed buildings and associated industrial activity ranging from about the fifth to eighth centuries AD (Sturgess 2007b), but the contemporary domestic settlement has not yet been located.

Phase 1		Blown sand deposits of the dune with occasional turf lines.
Phase 2	5th–7th centuries AD	Industrial features and pits.
Phase 3	7th–8th centuries AD	Buildings and a workshop complex with at least nine buildings (Figure 23).
Phase 4		Abandonment.

Two field systems were excavated in 1961 and 1963 (Fowler and Thomas 1962; Sturgess 2004, 172). One was in the depression between the dune and the Bronze Age settlement; the other lay in in an area of gently sloping ground close the base of the steep scarp of Godrevy Towans, about 100m north-west of the dune. Both of these excavations concentrated on post-Roman levels, although the existence of Bronze Age activity was mentioned.

Figure 23. Buildings from phase 3 of the post-Roman activity. The walls were partly rebuilt by the excavators. (© J. V. S. Megaw, Gwithian Archive)

Early medieval

St. Gothian's Chapel, south of the Red River, is one of the few 'old' sites known prior to the commencement of Charles Thomas's excavations in the mid-twentieth century. There seems to be no record of its construction date, but it is presumed to be early medieval. It became overrun by blown sand some time prior to the thirteenth century when the present Parish Church, 500m to the south, was founded (Edmonds 1863; Thomas 1964a). Some remains were seen by Leland during his perambulations round Cornwall in about 1538 (Hearne 1769, vol. 3: 18). It was then 'lost' for nearly 300 years before rediscovery in 1827 by a farmer who converted it into a cowshed (Haslam 1846; Edmonds 1863; Thomas 1964a). At the present time only isolated stones are visible in the turf mound covering the ruins.

'Sandy Lane' midden, lying about 50m north west of Gwithian village, was exposed by winter gales in 1963 and subsequently excavated (Thomas 1964b); it contained many hundreds of pottery sherds dating to the tenth to twelfth centuries AD. This may have been the site of a settlement which was abandoned in the eleventh or early twelfth century and which was superseded by the establishment of the manor at Crane Godrevy shortly afterwards.

Later medieval

The valley north of the Red River seems to have been permanently abandoned after the eighth century with no later evidence of any activity close the sites of earlier occupation.

There is evidence of multi-period occupation at the site of Crane Godrevy, on the level summit of Godrevy Towans at an elevation of about 70m OD. The site commands extensive views both of the coastline and inland. Excavations under the direction of Charles Thomas took place intermittently from 1952–1969. The earliest archaeology is a round dating from the Romano-British period, although possibly commenced in the late Iron Age (Quinnell 2007b), but which was later abandoned. Re-occupation for reasonably long periods spanned the mid to late medieval periods. A sunken-walled structure was on this site by the late tenth or early eleventh century which was expanded and modified into a 'manorial complex' by the fourteenth century (Figure 24). The site was abandoned around the end of the sixteenth century and was a ruin by about 1750 (Thomas 1954–55, 1969; Davies 2006; Nowakowski and Thomas 2007).

Figure 24. Crane Godrevy medieval manor at completion of excavations. (Nowakowski 2007b, Figure 23)

Chapter III – Previous molluscan studies at Gwithian

Palaeoenvironmental studies were not part of the original 1960s excavations at Gwithian but several studies have been undertaken in subsequent years, although only one has been fully published (Spencer 1975). Some of these studies were within the area of known prehistoric or post-Roman activity (Figure 25) and are discussed in this chapter. More recent studies in the wider landscape are discussed in Chapter XII.

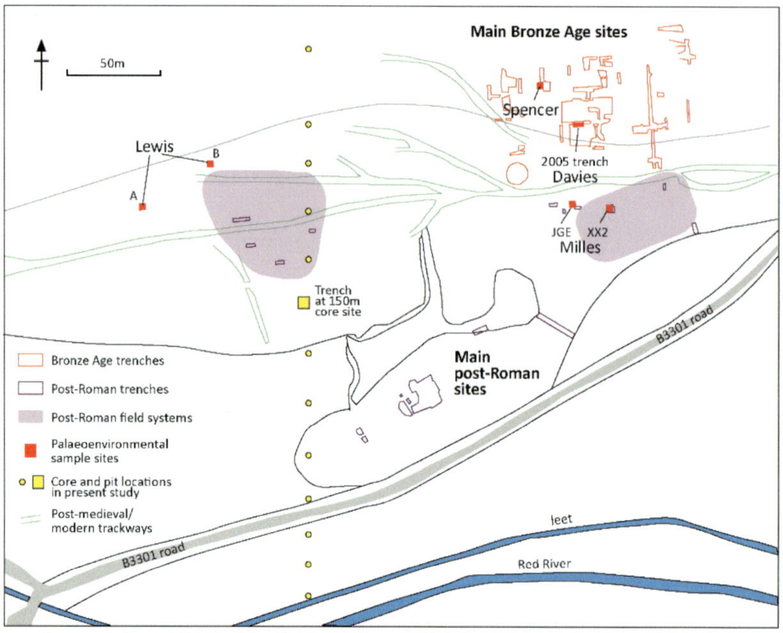

Figure 25. The excavations at Gwithian with the locations of palaeoenvironmental studies within the area of the Bronze Age settlement and post-Roman industrial sites.
The locations of the coring transect and trench in the present study are included.
(redrawn, with additions, from Nowakowski 2007b, Figure 2)

Lewis

The first study was by Geoffrey Lewis (1968) who investigated polymorphism of the shell *Cochlicella acuta*, but he also commented on the molluscs found in small trenches close to the post-Roman field system to the west of the Bronze Age settlement site. Full figures were only reported for one mollusc column (Profile B) and the mollusc diagram for this column is shown in Figure 26. There are, however, problems in interpretation of these findings. Large numbers of *Candidula intersecta* were found in lower levels, and in the light of present knowledge these are likely to be *Xerocrassa geyeri* (see below). Lewis did not count any helicids <5mm which introduces bias into his findings.

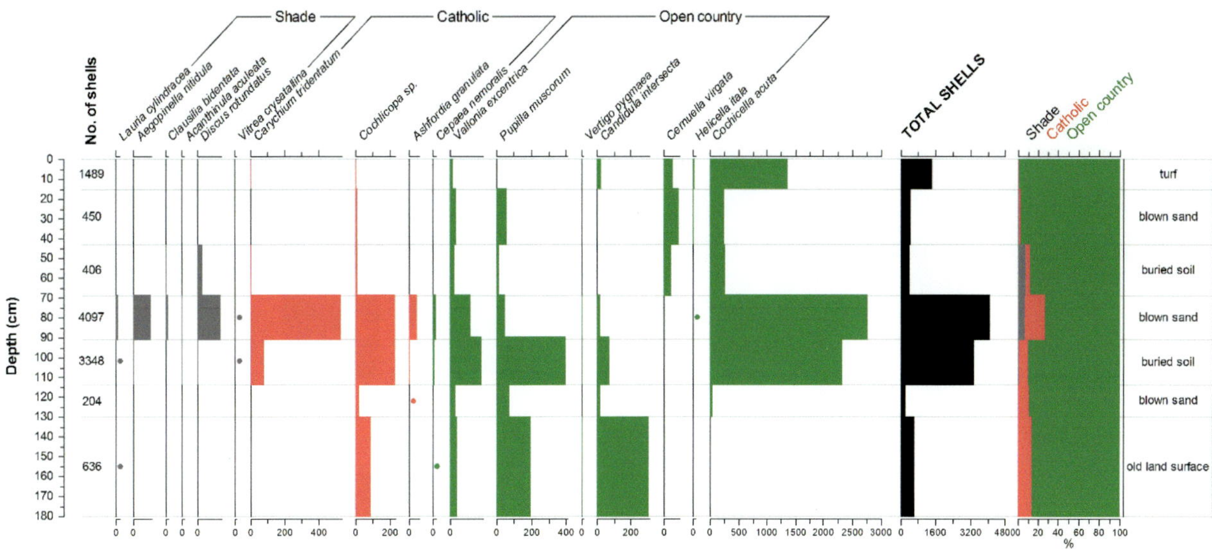

Figure 26. Mollusc diagram (absolute numbers) for the shells found by Lewis (Profile B). (note that the scale for *Cochlicella acuta* is different from that of the other species).

The stratigraphy was correlated with that of the Bronze Age site and Lewis ascribed the 43–69cm sediment 'with some confidence in the middle Bronze Age 2 (*c.* 1200–1000 BC)' (now regarded as late Bronze Age) and that at 91–115cm 'with less confidence in early Bronze Age or middle Bronze Age 1 (*c.* 1550–1350 BC)' (now regarded as middle Bronze Age). Palaeoenvironmental interpretation was very limited, but it is clear that while open country taxa predominate throughout the lower sequences, there is evidence of shade species in the middle levels (68–91cm) and which may represent an area of woodland or scrub; the location of his trenches is close to the base of the steep slope of the towans where scrub now grows and it is possible that this scrub extended onto the flatter valley floor in the Bronze Age, a habitat that would be suitable for *Discus rotundatus*. The higher sediments contain almost entirely open country species consistent with the meadow habitats that are now found in this area.

Spencer

Penny Spencer (1974, 13; 1975, 98) studied a mollusc column obtained from within the area of the Bronze Age settlement during the original excavations (Figure 27). Some shells in the Bronze Age deposits were named as '*Helicella caperata*' [*Candidula intersecta*] but were considered 'not typical' for this species; the shells were later examined by Milles (1991) who determined that they were *Trochoidea geyeri* [*Xerocrassa geyeri*]. The interface at 65cm depth, at the base of a plough soil horizon, contained plough marks cut into the underlying sand and has been radiocarbon dated to 1530–1010 cal BC (3070±103 BP, NPL-21) (Hamilton *et al.* 2008). The molluscan fauna is again mainly that of open country throughout the sequence but with some shade species in the Bronze Age sediments, most of which are able to live in rank grass. Early 'interference with the forest vegetation' was suggested as the reason why there was an absence of shade species at the base of the column, which overlay the underlying killas bedrock.

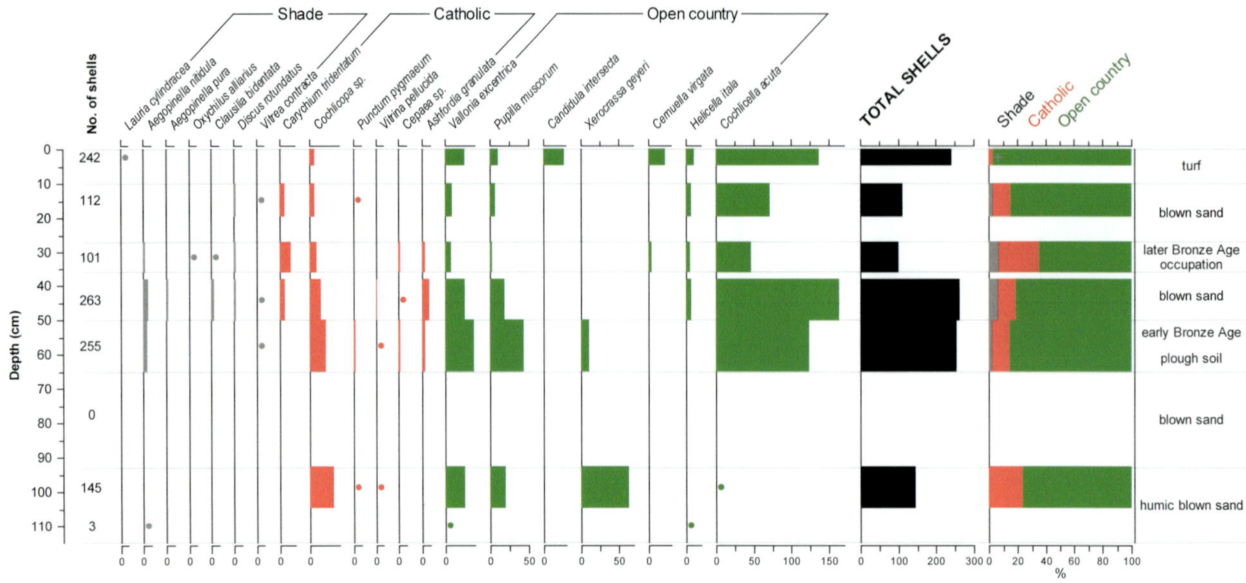

Figure 27. Mollusc diagram (absolute numbers) for the shells found by Spencer.

Milles

Some years later Annie Milles (1991) obtained material for mollusc analyses from small trenches opened immediately south of the main Bronze Age settlement in 1988 by John Evans. Column JGE (Figure 28) contained a buried soil at its base with very few shells and was probably non-calcareous. This was overlain by blown sand dominated by *Xerocrassa geyeri*, a xerophile typical of short open grassland. The plough soils from 85–100cm contained a richer molluscan fauna typical of dry grassland although there was a variety of species suggesting the presence of damper grassland. Blown sand covers the plough soil with reducing numbers of molluscs. Sediments above 75cm were not analysed, but consisted of blown sand up to the modern turf.

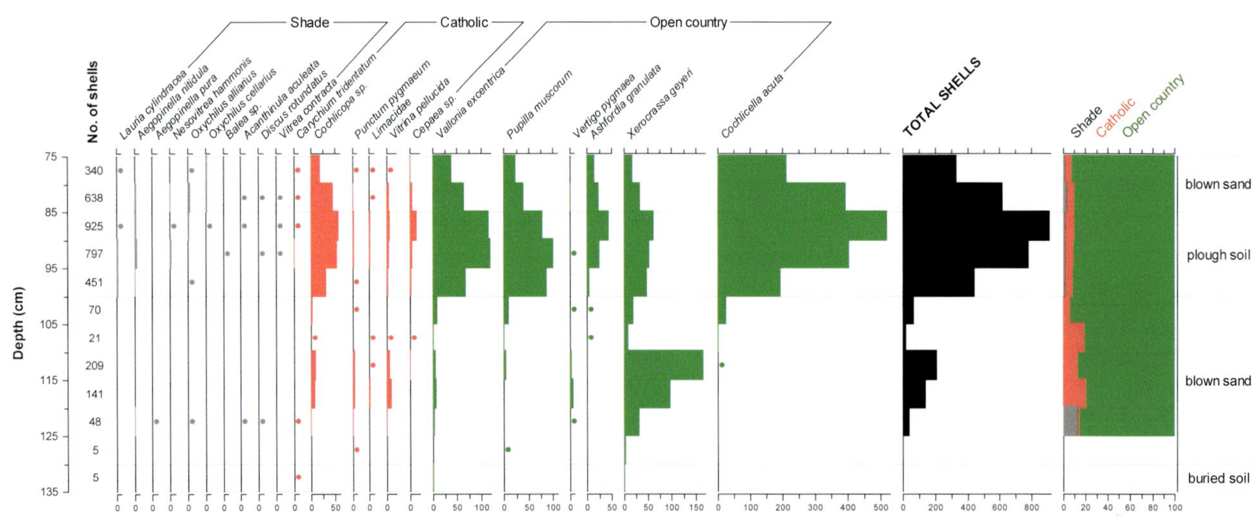

Figure 28. Mollusc diagram (absolute numbers) for the shells found by Milles in column JGE. (note that the scale for *Cochlicella acuta* is different from that of the other species).

The location of Milles' second mollusc column XX2 (Figure 29) was within a ridge and furrow field system containing plough marks which probably date to the post-Roman field period. The trench crossed a ditch cut into blown sand which overlies the Bronze Age sites, and which was included in the mollusc samples, the fill being from 67–87cm depth. This fill contains almost entirely open country taxa and is overlain by plough soil with a fauna consistent with short grazed grassland. It was considered that the restricted fauna was probably related to crop growing.

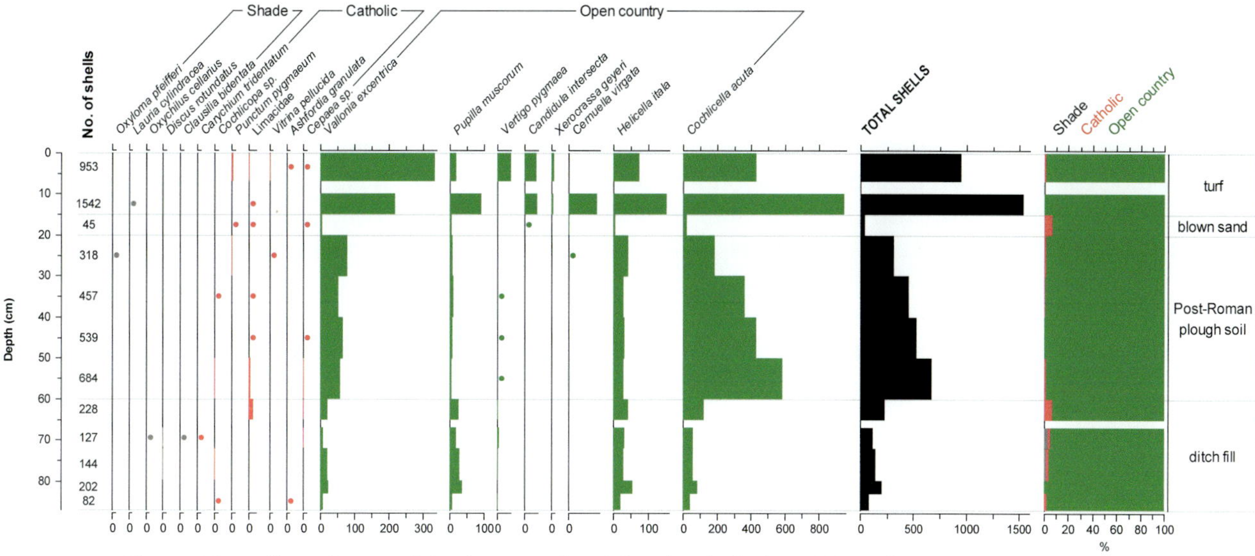

Figure 29. Mollusc diagram (absolute numbers) for the shells found by Milles in the area of the post-Roman field system (site XX2).
(note that the scale for *Cochlicella acuta* is different from that of the other species).

Davies

In 2005 a small trench was opened within the Bronze Age site at Gwithian to revisit the stratigraphic sequence and obtain samples for a wide variety of palaeoenvironmental studies (molluscs, pollen, charcoal, vertebrate bones, marine shells, lithics, pottery and stonework) and sampling for OSL dating (Nowakowski *et al.* 2006), but full analysis and reporting of these studies has not been possible because of withdrawal of funding from the Aggregates Levy Sustainability Fund. However, an assessment of the land snails was provided by Paul Davies (2006; 2007, 64) but it is not possible to construct a mollusc diagram as full mollusc numbers were not reported in the assessment. The deeper early and middle Bronze Age occupation levels with interspersed blown sand showed open country assemblages with a suggestion of later increasing shade; this was overlain by a high diversity assemblage, also middle Bronze Age, with a strong catholic and shade-requiring component and a few open country species implying a well-shaded stable environment, possibly woodland or scrub. A thin horizon above this may have been a dump deposit as it contained plant fragments, stony material, some marine shell fragments with a single freshwater *Pisidium* shell. The most superficial deposit, below the modern turf, again contained shade species, but with open country taxa becoming predominant in its upper part.

Discussion

These mollusc studies, performed either on shells obtained from the original excavations or from later trenches, all concentrated on the very local environments appertaining to the settlement sites or, in Lewis's study, a small area of the valley to the west. These show a continual open country environment with grazed grassland, although with suggestions of shade during phases of the middle and late Bronze Age. Whether there was scrub within the confines of the settlement, or merely areas of damper rank grassland, remains unclear. If woodland was present, then it was cleared prior to the initial settlement around 1800 cal BC. They provide no information on the palaeoenvironment of the wider landscape, and how changes in that landscape may have been affected by climate events. They give no indication of how settlement activity away from the main Bronze Age and post-Roman areas may have been influenced by climate variations. The present study is designed to explore the palaeoenvironment over a wide area around the excavated sites, both to the north and to the south of the Red River.

Chapter IV – The current study at Gwithian

INTRODUCTION

The current study was designed to realise the palaeoenvironmental potential of the Gwithian area, an archaeological site which helped lay the foundations of coastal and landscape archaeology. A mulitproxy approach has been used, with the main source of environmental evidence being mollusc analyses. The principal area of study relates to the archaeological areas described in the previous chapter, and comprises a coring transect along a line from the fields south of the Red River to the summit of Godrevy Towans at Crane Godrevy (Figure 30). There are no historical records of stream working in this part of the river which might disturb the stratigraphy on the valley; the 1877 Ordnance Survey 1st edition map does mark disused stream works 250m upstream from the present study area, although Stephens (1899) could find no evidence of working in this vicinity.

In addition to the coring transect a small trench excavation (2 x 2m) was undertaken on the north side of the Red River valley (Figure 30). This allowed more detailed sampling for palaeoenvironmental analysis, as well as an opportunity to visualize the stratigraphy in plan, albeit in a very limited area.

Figure 30. The transect line at Gwithian, showing the positions of each core and the excavation trench. The major archaeological areas excavated in the mid-twentieth century are also marked.

The environment over a wider area was investigated by obtaining a mollusc column from the western end of Godrevy Towans and from an area of cliff at the south end of Godrevy beach opposite Stack Rocks. These, together with other recent mollusc studies from the area, are discussed in Chapter XII.

FIELD METHODS

Multiple methods of investigation were used in the field, with fieldwork being carried out in June 2011 and between June and September 2012.

THE CORING TRANSECT LINE

The principal field technique used on the Gwithian archaeological site was coring. This method was selected as a long transect line would cause minimal damage to underlying deposits. From the outset it was assumed that the depth of sediments in the valley bed would be too deep or below the water table to permit excavation of a trench with the intention of reaching the basal natural geology, and this was confirmed when details of previous coring in the area were known.

The transect line was selected to avoid any settlement areas known from previous excavations and still allow a straight line from the valley to the towans summit close to the Crane Godrevy medieval manor. A line running north-south was initially plotted using ranging rods and a hand-held GPS unit (Garmin Etrex) which gave a claimed accuracy of 4–5m in both horizontal axes. At a later date the transect line was replotted using a Leica real time kinematic (RTK) differential GPS with a claimed accuracy of ±1cm in both horizontal and vertical axes (Figure 31); readings were obtained at 1m intervals along the entire transect except where vegetation or the river precluded access (Figure 32).

Figure 31. The differential GPS system in use on the summit of Godrevy Towans.

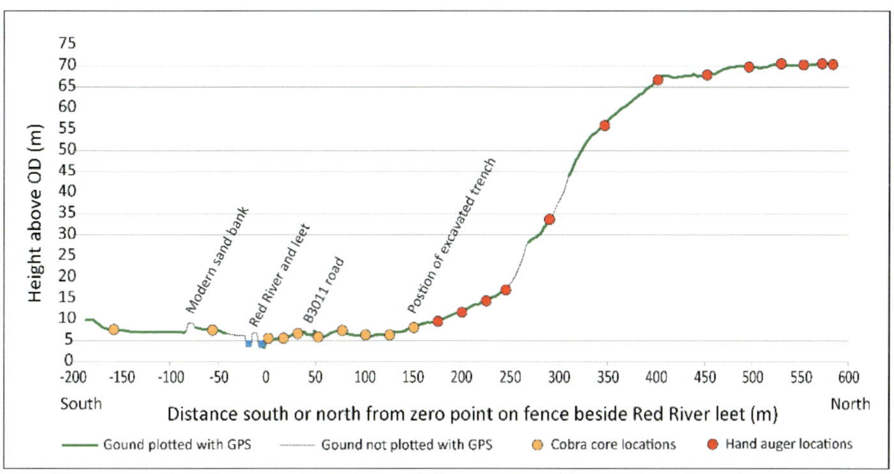

Figure 32. Graph showing the elevations of the transect line and the locations of the cores.

Ground penetrating radar (GPR)

Subsurface topography along the Gwithian transect line was examined using ground penetrating radar. It was considered that this might allow mapping of the stratigraphy between core locations, and help in correlation of the cores to the stratigraphy. GPR uses instruments which emit a radar signal into the ground which is reflected off structural interfaces in the sediments. The returning signals are recorded and stored in an on-site computer. The time between signal transmission and detection of the returning pulse is related to depth of the reflecting interface. When there are multiple interfaces, then the differences in transmission/receipt time allow a 'map' of the sub-surface stratigraphy to be plotted (for further information concerning GPR see David *et al.* 2008, 28).

The velocity of the radar pulse varies with the nature of the material through which the pulse is passing. Dry sediments such as sand or silt generally transmit radar waves at high velocity, but the presence of even a small amount of water, especially saline water or clay, considerably attenuates the velocity (Conyers 2004, 47). It can therefore be very difficult to determine reliable depth conversions when the sediments being studied contain a mixture of dry and wet materials, and the geological material comprising the sediments is variable. The water table at Gwithian varied between visits from 15-50cm below the ground surface close to the river indicating that much of the valley is permanently waterlogged. No attempts were therefore made to convert the time-based computer trace to an accurate depth-correlated trace using dialectric constants. However, a good estimation of depth could be obtained by direct comparison of the scans with the known depth of bedrock found on coring.

The depth to which radar waves penetrate depends on many other factors, one of which is the frequency of the transmitted pulse. Depth penetration is greater with lower radar frequencies, but the resolution of the resulting image is substantially reduced. If higher frequencies are used the resolution improves, but the depth achievable is correspondingly reduced. At Gwithian it was found appropriate to use probes of different frequency in different sections of the transect.

After establishing the line of the proposed transect the radar transmitter was dragged along the ground, the length of each traverse being limited by the length of the high-tension cable connecting the computer to the transmitter/receiver (Figure 33).

Figure 33. Ground penetrating radar.
Left: dragging the 200MHz probe through the vegetation immediately north of the Red River;
Right: the 100MHz probe in use in the field south of the river.

The GPR survey was conducted along the entire length of the transect except where vegetation or the river precluded access; the road running across the site was also not surveyed due to dangers from road traffic. Equipment used was Geophysical Survey Systems Inc. (GSSI) units. At the first radar visit in June 2011 scanning was performed with a 400MHz unit along the whole length of the transect from the Red River to Crane Godrevy and with a 200MHz unit from the river to the base of the scarp (245m) to obtain deeper penetration of the sediments. However, this frequency did not reach the base of the river valley and a return visit was made in September 2012 with a 100MHz probe. On this occasion GPR was performed north of the river for a distance of 150m and south of the river for 185m to cover the south side of the Red River valley.

Coring

Two methods of coring were used at Gwithian: percussion coring with a Eijkelkamp system driven by a Cobra percussor and hand coring with gouge augers. Percussion coring was used at those locations along the transect which were too deep to core satisfactorily with the hand auger as well as those in the lower part of the meadow. All the positions across the Red River valley and in the lower part of the meadow were cored in this way, while those in the upper part of the meadow, on the towans scarp and over the summit of the towans were cored using hand gouges.

The cores north of the Red River were spaced at 15m intervals near the river widening to 25m intervals in the meadow. Spacing on the scarp of the towans was wider, the exact positions being determined to some extent by access through scrub. Across the top of the towans the intervals were 50m reducing to 25m close to Crane Godrevy. The fence on the north side of the Red River leet was designated as the '0' point of the transect line. When the line was later extended south of the Red River the points, at wide intervals, were recorded with 'South' distances (Figure 30).

Percussion coring

The Eijkelkamp system uses a petrol driven percussion hammer (Cobra) to drive a metal tube vertically into the ground (Figure 34) (Canti and Meddens 1998). A 5cm diameter plastic lining tube is placed inside the metal casing; the lining tubes were opaque – an important consideration if OSL dating is required, as it is essential to prevent exposure of the sediment to light. At the nose end of the tubing a core catcher is inserted to prevent loss of sample on withdrawal of the tubes, which is held in place by a cutting head (Figure 35). The sample tubes are 1m in length with addition of 1m extension rods as necessary until the lowest desired depth is reached or until the nature of the sediment prevents any further insertion; at Gwithian it was found that the organic silt levels were very difficult to penetrate, probably because the silt, being spongy, reverberated so much with vibration that the corer moved up and down with each percussive action rather than penetrating deeper into the sediment. When this happened the sampling tube was replaced with gouge corers of diameters 3–6cm, the largest size being used when possible, reducing to smaller gauges when it proved impossible to insert the larger gauge gouge.

Figure 34. The percussion corer in use.
Left: the core sample tube being inserted into the ground. Right: the extraction process.

Figure 35. The lower end of the sampling tube.
From left: the cutting head; the core catcher; the lining tube which will contain the sediment sample; the outer metal casing. (photo: Eijkelkamp Soil & Water)

After one tube was fully inserted it was withdrawn and the inner lining tube containing the sample extracted and carefully labelled with the location code and core depth and the top and bottom indicated. The open ends of the tubes were sealed with cling film held in place with adhesive tape.

When the gouge corer was used the samples were examined in the gouge before removal. Boundaries between deposits were defined by eye and recorded. The depth of each deposit and Munsell numbers were recorded. The deposits were then removed from the gouge into pre-labelled plastic bags for subsequent analysis in the laboratory. If the deposit was up to 10cm in length it was removed in its entirety but if longer then it was divided into equal portions of no more than 10cm each.

Problems with percussion cores

There are difficulties with percussion coring, both as regards obtaining the samples and the quality of the samples obtained. Insertion of the sampler may be difficult if the vibration of the percussion unit is unable to advance the cutting edge, as mentioned above. At Gwithian the sediments are mainly sand or organic silt and while advancement through sand was generally steady the silt proved more difficult. Withdrawal of the sample chamber may also be extremely difficult with the manual extraction apparatus, illustrated in Figure 34. The pressure of sand binds on the sides of the chamber and considerable effort is sometimes required to retrieve the sampler, especially from greater depth. A further problem is that sand grains may be forced between the outer metal tube of the sampler and the inner plastic sample chamber. These are a tight fit and if grains become lodged between the two then it can be very difficult to extract the liner from the outer sheath.

The quality of the samples is affected in two major ways. This coring system does not use any form of sheath which remains in place after withdrawal of each 1m length, which means that after each extraction the core hole may become partially filled with sediment from the sides of the cavity. When the sampler is next introduced this extraneous sediment enters the sample chamber; although some will leave the chamber through an opening at its upper end much material may remain within the chamber. If the 'new' sediment is very compressible then more 'contaminated' sediment will remain in the upper portion of the chamber. While this may be clearly evident on opening the tubes if there is obvious mixed or inappropriate sediment in the upper part of the core it can sometimes be very difficult to determine the junction between the 'new' and 'contaminated' sediments. A method for calculating the true thickness of different sediments, allowing for compression and contamination, is discussed below.

A serious problem arises with percussion coring in the compression of samples. While this may be of minor concern with biological material that is flat or relatively incompressible, such as many plant stems, seeds and pollen, it can have major consequences for fragile items such as molluscs. The repeated actions of the compressor progressively fracture the shells and it is likely that some may become so fragmented that they become unrecognisable during the sorting process, which may lead to underestimation of both species and specimen numbers. Figure 36 shows examples of shells which have clearly been broken round their whorls into numerous fragments and are only held together by soil. These were extracted from unsieved sediments and the apices may well have been lost if subjected to wet sieving as is the norm in mollusc sample preparation.

Figure 36. Shells showing fracturing caused by percussion coring.
The graph paper has 2mm divisions.

Hand auger coring

Coring using a gouge auger was performed at Gwithian on the shallower parts of the transect where it was anticipated that the underlying natural could be reached without too much difficulty. A 2.5cm diameter gouge corer (1m length) was used at the initial visit. The quantity of material obtained using this gauge of core is limited and all the hand cores except those at 287m and 400m were later repeated using a 5cm diameter corer (50cm length). Unlike percussion coring there was no obvious damage to mollusc shells caused by the hammering action required to drive the augers.

The stratigraphy and Munsell number of each core was recorded on withdrawal from the ground. The sediments were then placed in labelled plastic bags, either as complete stratigraphic samples or with long sediments divided into portions, usually no more than 10cm in length.

Mollusc column samples

Samples for mollusc analysis were obtained taken in a standard way (Evans 1972, 41; Allen 2017, 33). Stratigraphic boundaries were respected in deciding where to divide samples. It is normal when taking samples from a mollusc column to commence at the lowest level to reduce the risk of contamination from falling material from higher levels. This was not possible on the sand dune sites due to the instability of the sand and all samples were taken from the ground surface downwards; care was taken to ensure that after one sample was taken the upper surface of the next sample was cleaned of any extraneous matter as much as possible to avoid contamination. Munsell colours were recorded on site while the sediments were still damp.

Trench excavation

After the cores were obtained and initial analyses performed it became evident that additional palaeoenvironmental information would be obtained if a small trench was excavated at one site along the transect line. This would not be possible in the bottom of the valley due to the depth of the deposits and it was decided that a site close to the known post-Roman field system in the meadow would be most suitable. The 150m location was selected as this lay between the southern limit of the field system excavated in 1960–62 and the margin of the presumed tidal estuary. The cores obtained previously had indicated that the underlying killas lay at a depth of about 100cm below ground level and that there was at least one layer of buried soil at a depth of about 60–70cm. The excavation is described in Chapter X.

Modern molluscs

Samples of turf were taken from each of the Gwithian core sites to assess molluscs which were present in very recent times. These samples provided modern analogues for conditions that may have been present in earlier times. However, the number of living animals (or those with intact periostracum) was found to be low in all of the samples, and it is accepted that there may have been some reworking of shells into the topsoil contexts.

At each site, a square approximately 20 x 20cm in size and 4–5cm deep was cut using an archaeological trowel and placed in labelled plastic bags. Photographs of each location were taken so that the mollusc assemblages could be matched as far as possible to current microhabitat preferences.

One sample was obtained from the Red River leet (at the '0m' location on the coring transect line) by dipping with a scoop with a 0.5mm mesh into the sediment at the base of the leet. Approximately 500ml of sediment was retrieved from which the molluscs were extracted. It is emphasised that mollusc numbers for this sample are not quantitative in regard to weight or volume of sediment; the sample is, nevertheless, valuable in assessing the current freshwater molluscs likely to be found in the area.

LABORATORY METHODS

From the outset, it was considered appropriate that a multiproxy approach be undertaken to gain a wide understanding of the sedimentary processes at Gwithian. Although mollusc analyses were the primary method employed, the findings were placed in context by performing particle size analysis of the sediments, as well as determining the pH, organic matter content and carbonate content of each deposit. Diatom and pollen analyses provided an alternative modality of palaeoenvironmental study, particularly at depths where molluscs had not survived. Geochemical analysis was used to explore the history of mining in the Red River catchment to be investigated. Micromorphology on samples from the trench was performed, principally to determine whether markings in the sand were due to ploughing. Dating of multiple contexts was obtained both by radiocarbon and optical stimulated luminescence techniques.

CORE SAMPLES

Percussion core samples

The cores containing the percussion samples consist of a plastic tube which must first be divided lengthwise into two halves. This was done using a circular saw set to a depth which just penetrates the thickness of the plastic without damaging the contained sediment. The tubes are then wrapped in cling film to keep them together and prevent dehydration.

When required the tube is unwrapped and the sediment divided into two halves by cutting lengthwise with a sharp knife and the two halves spread side by side. Once displayed it is necessary to decide how much contamination there is in the upper part of the core from material which has fallen into the bore hole between removing one length of core and inserting the next. This is usually fairly clear, as contaminating material frequently has an uneven mix of sediments of varying texture and colour and is different from new material in the lower part of the core. On occasions, especially in the sand cores at Gwithian, it was very difficult to determine the change from contamination to new sediment and a 'best guess' was made. In practice it was found that if the core was purely sand then 5–15cm was contaminant, whereas if there were thick silt layers then up to 65–70cm may be contaminant with subsequent marked compression of the silt (Figure 37).

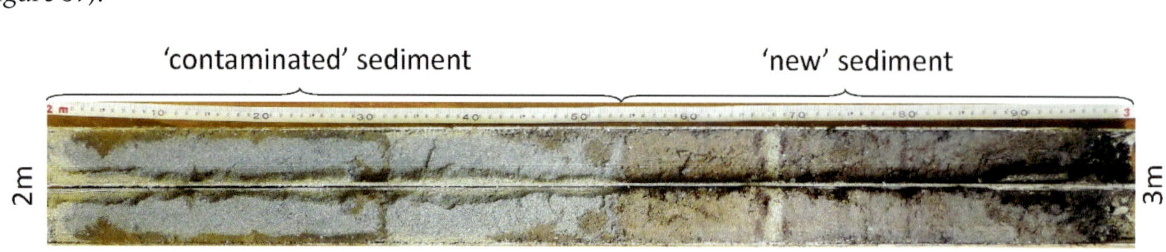

Figure 37. An example of contamination of a percussion core.
The 2–3m core from the 0.5m location, demonstrating contamination in the upper 54cm of the core.

The stratigraphic levels in a newly opened length of core were determined by inspection and recorded on a pro-forma, together with the nature and texture of the sediments, the abruptness of the interfaces, the presence of stones or other inclusions and the Munsell colours. One half of the core was then re-wrapped in cling film and stored as a potential archive for tests at a later date; however, it soon became clear that both halves of the core were necessary to maximise the number of molluscs in each sub-sample and all the results that follow include the whole core content. The sediments were removed into small labelled containers in lengths determined by the stratigraphy, with thick contexts being divided into portions. The sandy sediments were air dried and weighed. In the case of organic silt the material was kept moist and the whole examined microscopically for molluscs. Sieving was not used, and molluscs were extracted from the whole of each sample.

Correction method for compression caused by percussion coring

The problems with compression caused by the vibrations of the percussion corer have been mentioned above, and it was necessary to find a method to provide corrected depths for each stratum within the core. If the core consists of similar sediment throughout its length, such as all sand or all organic silt/peat, then conversion is easy, as, once the length of contamination in the upper part of the core tube is deducted, then the remaining length can be extrapolated to cover the full tube length (Canti and Meddens 1998). However, if there is a mixture of organic sediments in the core there will be different degrees of compression, and a straightforward conversion cannot be applied. Sand compresses relatively little whereas organic silt was subject to marked compression, with up to 70cm of contamination in some core lengths.

A method has been devised which to a large degree compensates for this. In several of the 1m core lengths in the present study the contained sediments were entirely sand, as at the 76m core location. Compression was found to be fairly consistent, with 5–15cm of contamination in each tube, with an average of 10cm per 1m length. Therefore, to obtain the correct thickness of each sand layer in the tube the measured thickness should be multiplied by a factor of 1.1 (there is, of course, no contamination in the 0–1m core, as this starts at the ground surface).

In a sequence of cores with differing sediments the following method was applied:

1. Add the measured total of all the sand deposits (but excluding 0–1m), after deducting the contamination lengths.
2. Multiply that total by 1.1 to give a corrected total for sand; this is also done for each individual sand layer within the cores.
3. If bedrock has been reached, then the difference between the corrected sand depth and the depth of the bedrock is the thickness which must be accounted for by organic silt.
4. Calculate the factor by which the silt/peat must be multiplied to convert the measured thickness to the difference found in (3) and apply this to each organic silt layer.
5. The total of the corrected thickness for sand and organic silt should equate to the depth of the bedrock.

This method will not be applicable if the bedrock has not been reached as the depth to which the lowest levels of silt have been pushed by the percussion process is not known. In practice when the bedrock has been reached the average compression ratio has been found to be 1.9–2.1 (i.e. the silt has been compressed to around half its 'true' thickness). A factor of 2 can therefore be applied in these circumstances and will provide corrected depths of acceptable accuracy.

Hand auger samples

As with the percussion samples the quantity of material of each subsample in the auger cores was small and wet sieving was not used and the whole of each sample analysed. There was no need to make any compression corrections, as none of the hand auger cores passed through organic silt, all being entirely sandy.

Mollusc analysis

Mollusc extraction by naked eye from sediments is inadequate for palaeoenvironmental purposes as the majority of very small shells and shell apices are easily overlooked (Sparks 1960, 1961; Allen 2017, 31). Wet sieving is therefore normally used with the finest mesh size being 500µm; this size has been found to give acceptable results (Evans 1972, 14), although it is probable that some very small shell apices (e.g. *Vertigo* spp.) may still be lost.

Core samples – percussion and hand auger

None of the samples from any of the cores was wet sieved. For many of the samples the total quantity of sediment in each sample was required for other tests (in particular particle size analysis and loss on ignition) and all material <500μm is lost in the process of wet sieving. All the sand samples were air dried and weighed and shells extracted under low power magnification using a stereo zoom microscope (GT Vision XTL-103BT with x0.7–4.0 zoom). All complete shells and all shell apices were removed and placed in a Petri dish, together with other relevant material such as charcoal, seeds, wood fragments. Non-apical shell fragments were only removed if there were few or no apices of the relevant shell so that the presence of that species could be recorded. Apices were the prime method of recording to ensure the 'minimum number of individuals', thus obviating duplicate counting (Evans 1972, 44). Almost all the shells were in remarkably good condition and further cleaning was not necessary.

The samples consisting of organic silt were partially dried and weighed. The whole sample was examined microscopically to extract the shells; it was found that with gentle teasing of the sediment while still moist, using fine forceps, it was possible to see and remove all the shells.

Mollusc column samples

Wet sieving was used for the samples obtained from the mollusc columns from the Gwithian trench, the modern Gwithian turf samples and the Strap Rock samples. The method employed is similar to that described by Evans (1972, 44). The samples were initially air dried. A suitable portion for mollusc analysis was weighed on domestic scales accurate to 1g. The weight used was determined by the likelihood of obtaining sufficient molluscs for reliable analysis and varied from 500g to 2kg. The appropriate weight to be sieved should preferably yield at least 150–200 shells per sample as this number has been shown to provide reasonable comparison of species (Evans 1972, 83).

Extraction of shells from bulk samples

The weighed samples were placed in a plastic container which was filled with water, and the sand gently stirred to ensure mixing. The supernatant was poured through a 500μm sieve to recover any shells and other material that floated (the 'flot'). This procedure was repeated several times until no more shells were released from the sediment.

The remaining sediment was then wet sieved through a stack of graded sieves of 2mm, 1mm and 500μm (it was not necessary to use a 4mm sieve due to the fine nature of the sands). The material passing through the 500μm sieve was discarded. Each sieve was then air dried, and, when dry, the molluscs and other material were carefully tipped onto paper, and then placed in plastic bags. Meticulous attention was paid at each stage to the labelling of sieves and bags. Extraction of shells from the dried sieved samples followed the same procedure as described above.

Identification of shells

Shells were sorted into individual species whenever possible, otherwise to genus. The number of specimens of each species present in each sample was recorded and the shells placed in labelled glass tubes or hard gelatine capsules. Identifications were made using standard texts (Evans 1972; Macan 1977; Kerney and Cameron 1979; Killeen *et al.* 2004; Cameron 2008) and a reference collection. Names applied are those published by Anderson (2005); the only exceptions are *Cochlicopa* cf. *lubrica*, *Cochlicopa* cf. *lubricella* and *Vallonia* cf. *excentrica* where, for ease of reading, the 'cf.' has been omitted.

For one particular species, *Xerocrassa geyeri*, now extinct in Britain, outside assistance was obtained. The shells were shown to Dr Richard Preece (University Museum of Zoology, University of Cambridge) who considered identification probable, but positive confirmation was received from Dr Edmund Gittenberger

(Professor of Systematic Zoology, Leiden University, Netherlands), an international expert on the Helicidae, the family to which this species belongs. Dr Gittenberger's opinion was that 'The shells are very well preserved and **very typical *geyeri* indeed**, i.e. in sculpture, shape (globularity), roundish very open umbilicus and size' (E. Gittenberger, pers. comm.).

Mollusc diagrams and diversity

Once identified the species were divided into different groups depending in their accepted habitat preferences, these being taken from published sources (Boycott 1934, 1936; Sparks 1961; Evans 1972; Davies 2008). Problems with using modern preferences as analogues for subfossil assemblages have been discussed above and it must be borne in mind that there will inevitably be some inaccuracies in allocation to particular groups. This is to some degree ameliorated by reference to the assemblages obtained from the modern turf at Gwithian, although this assumes uniformitarianism with ancient assemblages.

The ecological groupings used are: a) marsh/wet ground, b) shade, c) catholic and d) open country. The difficulties in using these ecological groups too rigidly has been discussed in Chapter I and will not be repeated here.

Mollusc diagrams were constructed using the software C2, version 1.7.6 (Juggins 2011). Diagrams for absolute numbers of shells have mainly been used throughout as for many samples the number of shells was too small to be meaningful on a percentage basis. Diagrams showing the relative numbers of shells have been constructed when appropriate. For all the mollusc column samples the actual number of shells found in each weighed sample was used. However, for the core samples the sample weights varied markedly, and diagrams have been constructed with mollusc numbers normalied to 500g samples in each case.

One requirement of mollusc analyses is that samples should contain a mollusc assemblage within the sample that is representative of the overall population. Species diversity indices can provide some indication of adequacy (Law 2017) and was calculated in this study using the Shannon (Shannon and Weaver 1949) and Brillouin (Brillouin 1956) indices, calculated using PAST software (Hammer *et al.* 2001). The difference between these two indices is an indicator of whether that sample is adequately representative of the overall assemblage (Peet 1975; Allen 2009).

Evans (1972, 83) has demonstrated that 150–200 shells in a sample 'is a sufficient number to establish the broad composition of a fauna'. The majority of core samples and some column samples fail under this criterion. To test sample adequacy the difference between the Shannon and Brillouin indices against number of shells was calculated for every sample analysed in the present study. 163 mollusc column samples and 106 core samples were assessed (Figure 38) but samples with less than twenty-five shells are not shown on the graph. It is clear the greater the number of shells then the nearer the two indices approach each other.

A Shannon-Brillouin difference of 0.1 seems to be a 'break' point, and if this were taken as an acceptable level then only two samples with more than 180 shells would fall above this figure, entirely in accord with Evans's experiment. It is evident that most of the core samples fail this criterion, and even if a Shannon-Brillouin difference of 0.2 was accepted, then many samples would still be inadequate. In the present study it is clear that the majority of the core samples and many of the mollusc column samples may not represent the complete assemblages from which they are derived. Despite these limitations (which apply to many mollusc studies) it is felt that analyses of samples with fewer than 180 shells can validly be used as a guide to habitat types.

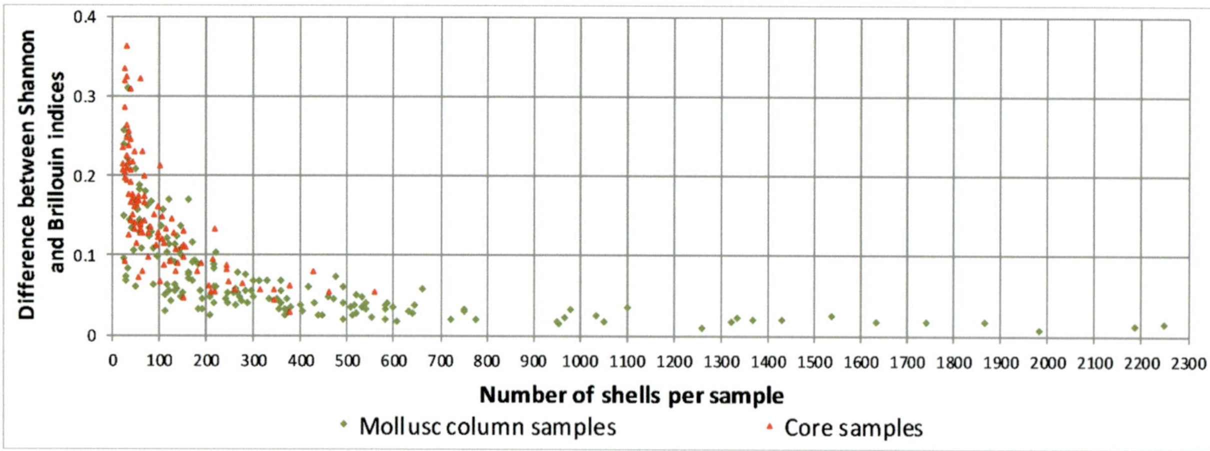

Figure 38. Comparison of shells per sample with difference between Shannon and Brillouin diversity indices.

Particle size analysis

Particle size analysis was performed using the Coulter™ LS 230 laser granulometer (Pye and Blott 2004). Particles in the size range 0.4µm to 2mm are measured using a laser system and from 0.04µm to 0.4µm with polarisation intensity differential scattering (PIDS). Sand samples from the cores for particle size analysis were taken from the subsamples after mollusc extraction. For the bulk samples from other sites material which had been passed through a sieve with 2mm mesh was used as the granulometer cannot accommodate particles larger than 2mm. Approximately 40g of each sample was repeatedly passed through a riffler until a sample of 1–3g. remained; this was disaggregated with Calgon (sodium hexametaphosphate) at a dilution of 3.3%. For the silt samples a quantity of sediment was smeared on a palette with water to make a paste and small subsamples taken from different areas to give a representative sample prior to disaggregation with Calgon. It was not considered necessary to remove organic particles prior to analysis (Allen and Thornley 2004). The sample was then poured into the granulometer chamber and measurements performed.

Analysis of the raw data was performed using the Gradistat software programme (Blott and Pye 2001). Sediment statistics and graphs of class volume *vs* particle size were used to compare the composition of the various samples in relation to depth below the present ground surface.

pH estimation

The pH of samples was measured using a Hanna Instruments pH209 meter which was calibrated with pH 4.00 and pH 7.00 buffers for acid sediments and pH 7.00 and pH 9.22 for alkaline sediments.

The method outlined by Rowell (1994, 160) was used. 10g of air dried sediment was weighed into a 50ml centrifuge tube and 25ml of ultra-pure water added; for a few samples there was little material and 5g was used with corresponding reduction in the quantity of water. The tubes were then placed in a rotating shaker unit for 15 minutes. The pH was measured using the meter probe which was rinsed with ultra-pure water between each measurement. Two measurements were made for every sample with a third measurement if the first two differed by more than 0.2 units. The mean of the two/three measurements was then calculated.

Loss on ignition

The organic matter content and carbonate content of samples was determined by loss on ignition (e.g. Ball 1964; Dean 1974; Howard and Howard 1990; Heiri *et al.* 2001; Santisteban *et al.* 2004). Recommended times and temperatures vary between authors but the principles are the same. A sample is first completely dried by placing it in an oven at 105°C and then all organic matter is burned off at 450–500°C; this is followed by heating to 850–1000°C when carbonates are oxidised. The times and temperatures used in this project are those which are standard in the Department of Soil Science, University of Reading; dehydration was achieved by 24 hours at 105°C, followed by 24 hours at 500°C to determine organic matter content and finally two hours at 950°C for carbonate content. At each stage weights were measured to four decimal places. The carbonate content was multiplied by a factor of 1.36 to allow for the loss of some carbonate during the heating at 500°C. Using this factor shows good correlation with other methods of determining carbonate content (Heiri *et al.* 2001).

X-ray fluorescence (XRF) geochemistry

When materials are exposed to X-rays the component elements undergo ionisation, and in the process radiation is emitted which is characteristic of individual atoms; i.e. the atoms 'fluoresce'. This fluorescence can be detected and measured quantitatively, either as differing photon energies or by separating the emerging wavelengths.

Itrax fluorescence of cores

An automated XRF method of obtaining geochemical information from core samples has recently been introduced (Croudace *et al.* 2006). In the Itrax (Cox Analytical Systems, Gothenburg, Sweden) system the half-section of the split core is placed in a cradle which then passes incrementally beneath the detector housing (Figure 39). It gathers optical and radiographic images as well as XRF elemental and magnetic susceptibility profiles. The surface of the core to be scanned must be fairly smooth with no stones or other material protruding above the surface more than about 2–3mm. XRF scanning resolution is practicable at intervals of ≥100μm. The X-ray tube anode can be either molybdenum (for elements with a higher range of atomic number) or copper (for a lower range of atomic number). Unlike pellet XRF which measures quantities of elements present, Itrax scanning measures the counts emitted by each element over a pre-determined length of time.

A molybdenum anode was used at 30kV and 35mA with a data acquisition time of 10 seconds and with scanning increments of 1mm for the XRF and 5mm for magnetic susceptibility. For analysis of the geochemistry running averages of 2cm were used to provide some smoothing of the curves. Itrax scanning for this study was performed by the author in the Department of Geography and Earth Sciences in Aberystwyth University under the guidance of Dr Sarah Davies.

Figure 39. The Itrax core scanner. The core passes from the cradle holder on the left, beneath the detector array in the centre, to the receiving cradle on the right.

Micromorphology

Micromorphology on two samples from the trench in the Gwithian meadow was performed by Dr Rowena Banerjea (University of Reading). Full details of the technique used are given in Chapter XI.

Pollen analysis

Ten samples from the organic silt layers in the 0.5m core at Gwithian were initially examined to see if pollen was preserved in sufficient quantities for analysis. These proved rich in pollen and a further 55 samples taken from the core to cover the entire length of silty sediment throughout the core. Ten of these were prepared for analysis for the current study, with the remaining 45 samples placed in cold store for potential study at a later date.

The slides for analysis were prepared by Kevin Williams (University of Reading), using the method outlined in Branch *et al.* (2005, 126), with deflocculation in sodium pyrophosphate, acetolysis and separation of the minerogenic fraction with sodium polytungstate and with Erdtman's mixture; the slides were stained with safranin. The prepared slides were examined and pollen counted by Dr Rob Batchelor (University of Reading), whose full report is included in Chapter VIII.

Diatom analysis

Slides for diatom analysis from six samples from the 15m core at Gwithian were prepared by Kevin Williams (University of Reading), using the method outlined in Branch *et al.* (2005, 126). Preparation involved disaggregation with 30% hydrogen peroxide, digestion in 10% hydrochloric acid, centrifuging to concentrate the diatoms and then removal of clay particles with dilute ammonia, before mounting on microscope slides. The prepared slides were sent to Dr Nigel Cameron (University College London), who kindly examined them and reported on the species present using x400 and x1000 magnification. The aim of diatom analysis was to determine whether any marine species were present, and identifications only were performed. Counting of numbers present was not done.

Optically stimulated luminescence (OSL) dating

Two OSL dates had previously been obtained at Gwithian from sands in a trench opened in 2005 (Hamilton *et al.* 2008), and it was considered appropriate to obtain further OSL dates on the new material from the present study. The Aberystwyth Luminescence Research Laboratory, (ALRS), Aberystwyth University, was chosen as the 2005 samples had been dated in that laboratory; the new analyses were performed under the direction of Dr Helen Roberts. Details of OSL methods are given in Aitken (1994), and in reviews by Duller (2004, 2008) and Roberts (2008).

Five OSL samples were taken during excavation of the trench opened at Gwithian and a further five from the 30m core, subsampled at Aberystwyth. This core was selected for various reasons: it was the deepest core (10m), it had a good range of stratification of both sand and organic sediments, and one half of the split core was available to take the OSL sub-samples before it was exposed to white light. Environmental dose rates were assessed in the field for the trench samples and in the laboratory for the 30m core samples.

Measurements were made on coarse grained quartz, using the single aliquot regenerative dose (SAR) protocol of Murray and Wintle (2000, 2003).

Radiocarbon dating

Application for radiocarbon dating was made to the National Environmental Research Council (NERC), and a grant approved (NF/2103/14) for 19 samples to be dated. The dating was performed at the NERC Radiocarbon Facility (Archaeology) at the Oxford Radiocarbon Accelerator Unit. Organic matter was generally only present in the organic silt layers in the deeper core samples and many of the dates were obtained on land shells.

Eight of the samples were taken from the core obtained at 30m from the Red River; this core was considered best to use as five samples had already been selected for OSL dating. A sequence of 13 dates was thereby obtained from a single core sequence. Six radiocarbon dates from these samples were on alder wood (*Alnus glutinosa*), one on bog-bean seeds (*Menyanthus trifoliata*) and two on *Cochlicella acuta* shells.

Four samples for radiocarbon dating came from the trench in the Gwithian meadow, all from the two buried soil horizons and all on the shell *Xerocrassa geyeri*. These dates provide correlation with the OSL dates from the same trench.

Seven radiocarbon dates were obtained from samples of shells taken from the wider landscape. Three were from the exposure at Strap Rocks, (the uppermost *Cochlicella acuta*, the lower two *Xerocrassa geyeri*). This provided chronology for the newly discovered feature eroding from the cliffs, and also provided additional dating evidence concerning the extinction of *X. geyeri* and the introduction of *C. acuta*. A further four samples (all *Cochlicella acuta*) came from stabilisation layers identified in 2007 in the West Godrevy Towans samples, and provided dating of the deposition history at a site remote from the main occupation areas at Gwithian.

Standard methods were used for the radiocarbon dating. Isotopic fractionation was corrected for using the measures $\delta^{13}C$ values measured on the AMS. The quoted $\delta^{13}C$ values were measured independently on a stable isotope mass spectrometer (to ~0:3 per mil relative to VPDB). Details of the chemical pre-treatment, target preparation and AMS measurement are given in Bronk Ramsey *et al.* (2002; 2004). Calibrations were performed with OxCal 4.2 software using the IntCal 13 curves (Reimer *et al.* 2013).

Chapter V – The coring transect

TURF BLOCKS

The numbers of each species identified from the samples is shown Table 1 and graphically in Figure 40. A total of 3679 shells was identified in the 23 samples comprising 27 taxa. Only a very few of the molluscs (approximately 3%) were live at the time of collection, although more retained an intact periostracum or were translucent indicating recent death. The assemblages therefore do not accurately represent a modern living population, but are indicative of the population in the recent past when the very superficial sands were accumulating, and are a guide on how to interpret the palaeoassemblages.

Table 1. Number of molluscs in the modern turf samples. 500g samples.

Distance from Red River (m)	South 158	South 58	0.5	15	30	51	76	100	125	150	175	200	225	245	289	345	400	451	495	528	551	570	582
Habitat	bare field	bare field	lush, dense, vegetation	shorter vegetation	grass tussocks	in shadow of Cornish hedge	short grass dune	short grazed grass	short grazed grass	short grazed grass	short grazed grass	short grazed grass	short grazed grass	at edge of scrub	in area of dense scrub	very bare open hillside	marram grass	short grazed grass	at edge of scrub	in shadow of scrub	short grazed grass	short grazed grass	at edge of tall scrub
Carychium minimum					1	83			1														
Galba truncatula						1			1														
Cochlicopa lubrica					17	1						5	5	4	4	3	5	1	28				1
Cochlicopa lubricella															4								
Vertigo pygmaea					65		18	6	7	15	32	40	52	11	1	14	7	25	4	6	24	18	1
Pupilla muscorum							5			1											20	38	3
Lauria cylindracea					2										1	9	72	2	16		2		
Vallonia excentrica					36	51	52	27	117	62	157	78	216	144	2	36	9	151	25	28	133	22	65
Punctum pygmaeum					21		2			8	5	8	5	5		1	15	34	1	1	32	9	4
Vitrina pellucida					3	1	1				1									9			
Vitrea crystallina					24		3										4		1				
Aegopinella nitidula				1	2										5		39		7				3
Aegopinella pura			1																				
Nesovitrea hammonis						8																	6
Oxychilus alliarius					8							9							22	28			
Milacidae						1																	
Limacidae									4		2	2	3	1					1		1		
Cecilioides acicula																			4	1			
Clausilia bidentata					2											1	1						
Cochlicella acuta				1	50	4	26	26	90	27	23	12	66	95	14	125	22	27	6		8	11	1
Ashfordia granulata					19				1		1		1				3		1				
Candidula intersecta					10		6	4	1	1			1		2	23	12	8	3		4	26	
Cernuella virgata			1		18		4	4	8	3	5		6	17	1	34					2	1	
Helicella itala				1	8		15	30	54	10	8	4	17	22	4	47	7	23	9		3	21	5
Trochulus hispidus						17	1		15	1	3		1		3	5	2	1	3	1			1
Cepaea nemoralis			2		3	1						2				1	3	1	4				
Cornu aspersum						1									1								
TOTAL SHELLS	**0**	**0**	**4**	**3**	**289**	**169**	**133**	**97**	**299**	**128**	**237**	**160**	**373**	**299**	**42**	**299**	**201**	**273**	**135**	**74**	**229**	**146**	**90**
TOTAL TAXA	**0**	**0**	**3**	**3**	**17**	**11**	**11**	**6**	**11**	**9**	**10**	**9**	**11**	**8**	**12**	**12**	**14**	**10**	**16**	**7**	**10**	**8**	**10**

There were no shells in the two turf blocks taken south of the river and very few in the 0.5m and 15m samples; possible reasons for this will be discussed below. At all other locations there was a good number of shells, but with considerable variation at different sites along the transect; understanding these variations aids the interpretation of subfossil assemblages. Figure 41 shows how the number of shells and taxa varied from site to site; generally, each rose and fell in accord, but there are discrepancies at locations 245m, 289m and 497m which will be discussed under each core. Only some general comments will follow here.

The molluscs found in the modern turf samples at Gwithian contain species from all ecological groupings, sometimes in locations where they might not normally be expected. The most obvious example is at 30m, an area of rank grass but with no scrub or woodland within about 200m; the only shade is among stems of grass so it is perhaps surprising to find two specimens of *Clausilia bidentata*. Other species normally associated with shade are *Vitrea crystallina*, *Aegopinella nitidula* and *Oxychilus alliarius*, but all of these may live in tall damp grassland, and *O. alliarius* may be one of the first colonisers of grassland left to become rank (Davies 2008, 177). If any of these species are found in sub-fossil assemblages then the presence of a good degree of shade is to be expected, although not necessarily woodland or scrub.

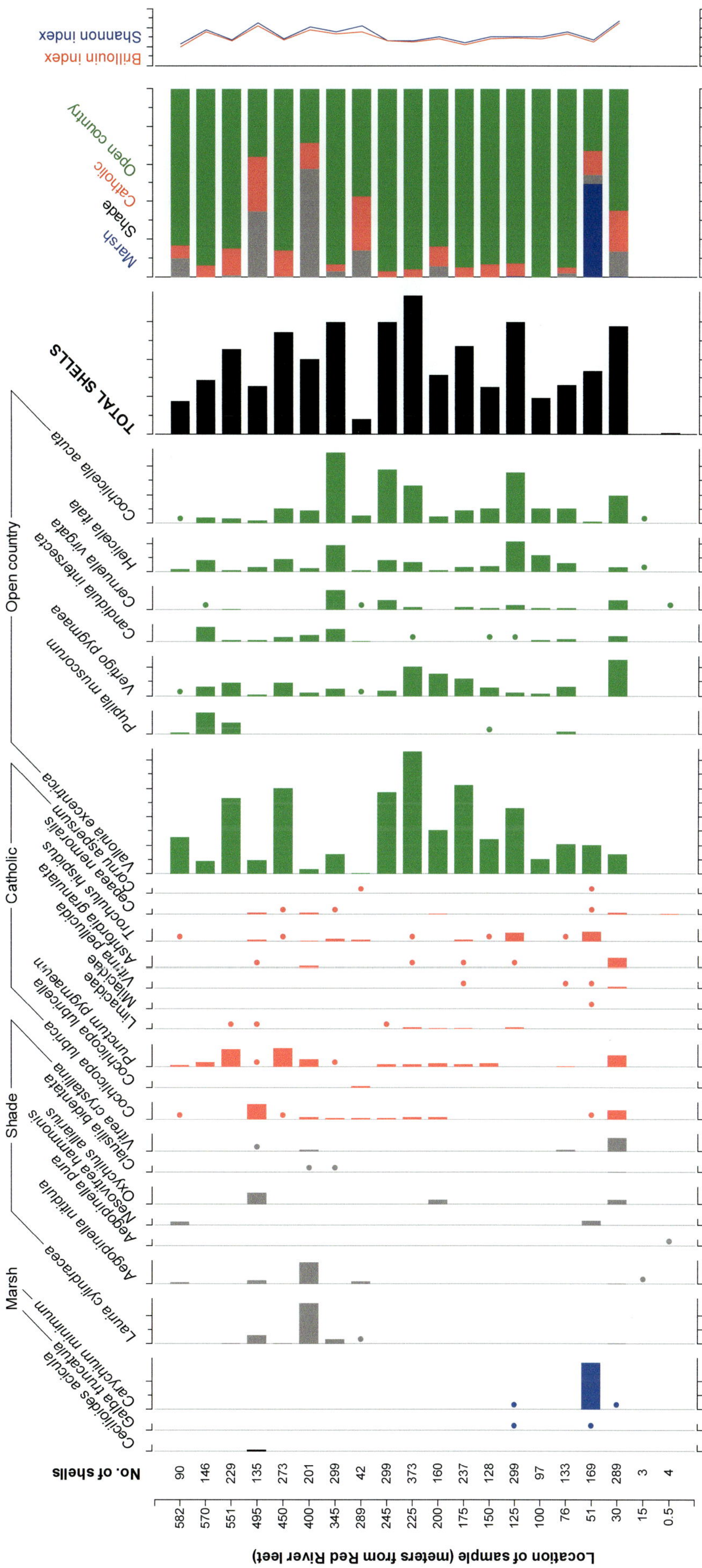

Figure 40. Mollusc diagram for the modern turf samples, arranged from south (bottom) to north (top). The two locations south of the river are not shown as neither contained any molluscs. Absolute numbers of molluscs. Dots indicate single specimen only present in sample.

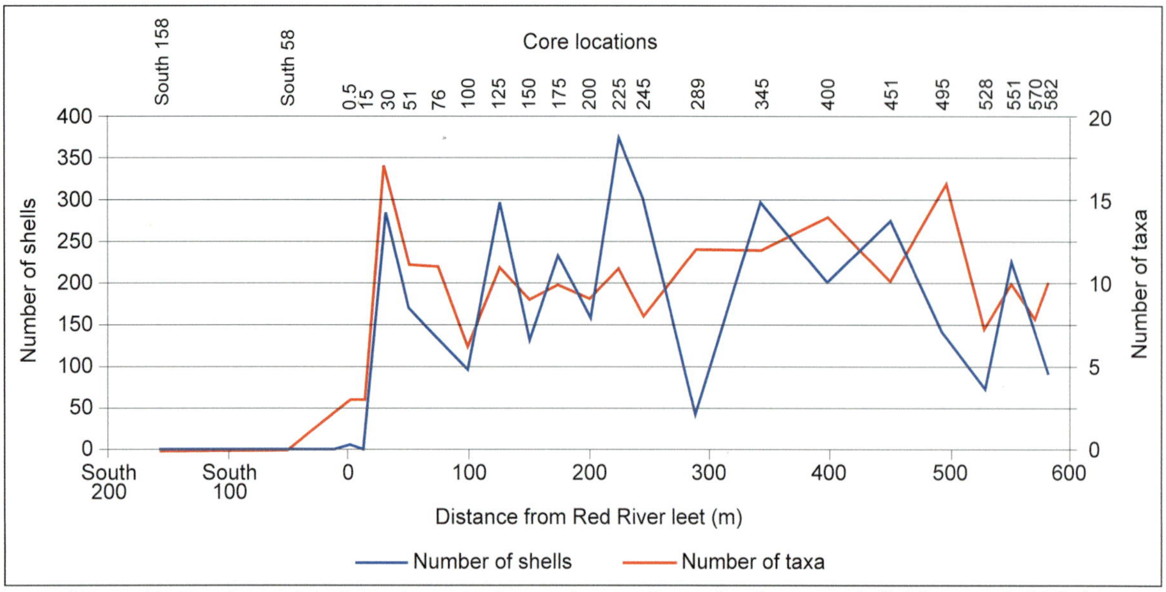

Figure 41. Numbers of shells and taxa in the modern turf samples, from south (left) to north (right).

The shady and damp habitat provided by the 'Cornish hedge' at the 51m position is also instructive (the construction of this type of 'hedge' is described in Chapter VI). This low-lying area is very damp or under water for much of the year, and the perpetual moisture is consistent with the presence of large numbers of *Carychium minimum* and a single *Galba truncatula*; *Nesovitrea hammonis* is also capable of living in wet meadows. However, the presence of many *Vallonia excentrica* is, perhaps, unexpected, although this species may occasionally live in marshy habitats (Evans 1972, 162).

The open country species found on the meadow and towans summit and can be used as an analogy when present in subfossil assemblages, indicating short grassland which may be grazed. When in association with species which require some shade, such as *Aegopinella nitidula*, then it is likely that there was at least long rank vegetation, as at 400m, or scrub in the close vicinity, as at 289m and 495m. There is no true woodland at any site along the transect line (the nearest trees are about 700m to the east where there is a line bordering a track), so there are no modern analogues for woodland in the immediate area. If subfossil assemblages differ significantly from those found above and contain a higher proportion of shade species, then the presence of woodland should be considered but is not proven.

Specimens of flowering plants were collected within approximately a 2m radius of each core site in August 2011 and kindly identified by Steve Hebdige and Priscilla Oates. Details of the species found are given with the discussion of each core location.

Molluscs from the Red River leet

Approximately 500ml of sediment was obtained from the bed of the leet where it crossed the coring transect line and wet sieved down to 500µm. The pH of the water was 7.65. Very few molluscs were alive at the time of collecting, the majority of these being *Potamopyrgus antipodarum*. The shells recovered are listed in Table 2.

The majority of shells are catholic freshwater species (83%), those which require continuous wet conditions without periods of drying, but which may have been carried to this site from far upstream. The leet is generally slow flowing, which explains why there is only one moving water species. Of note is the presence of large numbers of *Potamopyrgus antipodarum*, a nineteenth century introduction to Britain which is ubiquitous in all types of flowing and brackish water and tolerant of mildly polluted waters. There is a wide range of terrestrial molluscs which have clearly been deposited here from further upstream where there are woods down to the water's edge.

It is important to observe that in the cores, the discussion of which follows, almost no true freshwater molluscs have been found. If there had been regular deposition of overbank alluvium then some at least of these would have been expected, and their absence considerably reduces the likelihood that overbank flooding occurred more than very infrequently.

Table 2. The molluscs found in the sample from the Red River leet.

FRESHWATER 565 shells	
MOVING WATER	
Ancylus fluviatilis	2
CATHOLIC	
Radix balthica	23
Gyraulus albus	71
Gyraulus laevis	102
Pisidium spp. (including *P. milium*, *P. nitidum*, *P. subtruncatum*)	282
DITCH/SLUM	
Potamopyrgus antipodarum	71
Physella acuta	14

WET GROUND 12 shells	
Galba truncatula	3
Carychium minimum	7
Anisus leucostoma	2

TERRESTRIAL 38 shells	
SHADE	
Lauria cylindracea	
Punctum pygmaeum	1
Discus rotundatus	1
Vitrea crystallina	1
Zonitoides nitidus	6
Euconulus alderi	2
Balea heydeni	2
CATHOLIC	
Cochlicopa lubrica	4
Cochlicopa lubricella	1
Cochlicopa sp.	4
Ashfordia granulata	4
Trochulus sp.	1
Cornu aspersum	4
OPEN COUNTRY	
Pupilla muscorum	1
Cernuella virgata	2
Cochlicella acuta	4

GROUND PENETRATING RADAR

The radar survey was conducted in several sections:

south of the Red River;

between the river and the base of the Godrevy Towans scarp;

the upper scarp and Towans summit, as vegetation allowed.

In no part of the main valley did the radar reach the bedrock, although this was visible on the slopes north of the river. The depths indicated on the 0.5m to 150m traces are extrapolated from the known depths of interfaces and bedrock observed in the cores. No attempt has been made to determine depths directly from the radar signals.

SOUTH OF THE RED RIVER

This section was scanned only with the 100MHz probe (Figure 42) but no clear interfaces were seen. While some of the subsurface lines may represent subtle changes in stratigraphy these were not evident in the cores and no interpretations have been attached to this section of the GPR scan. Of particular note is that the scan did not show any evidence of the valley slope to the south.

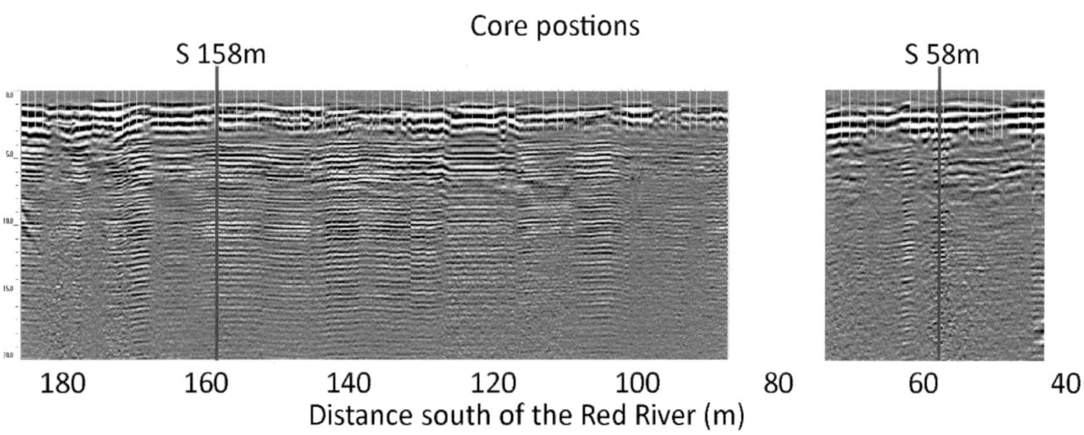

Figure 42. GPR scan (100MHz) of the field south of the Red River, with the core locations marked.

Between the Red River and the Towans scarp

Scans were done between the river and the 150m core location with the 200MHz and 100MHz probes, while those between the road and the scarp also included scans with the 400MHz probe. Multiple oblique interfaces are revealed between the river and 150m (Figure 43) which are outlined in Figure 44. Those from 50m to 150m are interpreted as the northern bank of the river valley gradually infilling with sand over time, while those from 0m to 50m are changes in the sediment layers sufficient to produce radar reflections. The 400MHz scans over this distance are not illustrated as they did not provide additional information. Good detail was obtained in some areas with the 200MHz probe to an extrapolated depth of about 3–4m, while the 100MHz revealed some interfaces to a depth of about 8m.

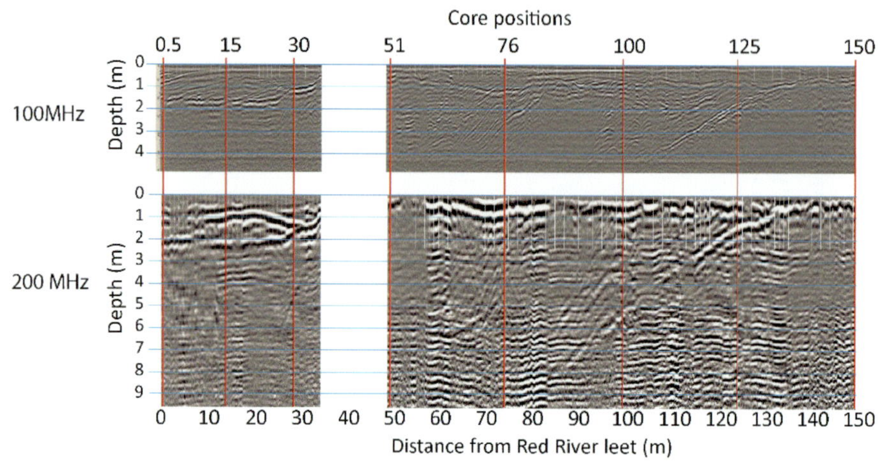

Figure 43. GPR between the Red River and the 150m core position, with the core locations marked. Depths extrapolated from known bedrock depths at 100m and 125m. The gap in the sequence is the B3301 road.

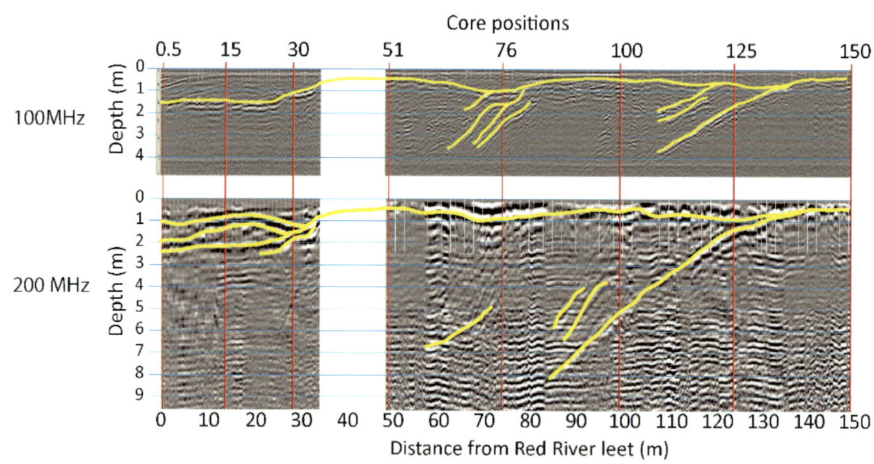

Figure 44. The strong radar interfaces drawn onto the GPR scans.

It is difficult to correlate many of these potential interfaces with the stratigraphy found in the cores, and which are described in more detail below. The 0.5m and 15m cores both show clear stratigraphic changes from sand to organic silt at 120–145cm, corresponding very well with the radar, but deeper boundaries at 600–700cm are not revealed in any area. The 30m core shows stratigraphic changes at 172cm and 207cm in accordance with the GPR. No horizons below 50cm are seen at the 51m location with the scan being very noisy, largely because of the difficulty encountered in maintaining contact of the radar probe with the ground due to grass tussocks. Several boundaries are suggested at 76m but none seen in the core, so these appearances are equivocal.

A sloping bank is clearly demonstrated between 85m and 135m on the GPR, and this corresponds with the bedrock levels found in the 100m and 125cm cores, and has allowed depth models to be applied to the GPR.

The GPR between 150m and 245m, shown with the 400MHz probe (Figure 45), reveals little detail which is perhaps only to be expected as the cores reached the killas at depths between 22cm and 100cm, with no significant stratigraphic horizons except some undulation on the possible bedrock interface, but this is equivocal. Of interest is that there are no features at all between 174m and 201m, corresponding closely to the position of the post-Roman field system; the disturbance of the ground in this area seems to have resulted in radar noise only with loss of all interfaces.

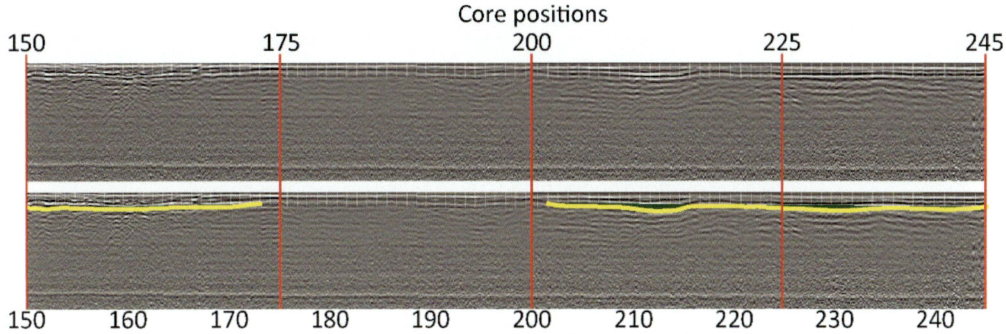

Figure 45. GPR between 150m and 245m (400MHz).
The lower image shows the interface between the bedrock and overlying sandy sediments drawn onto the same scans.

The Towans

It was not possible to perform any GPR on the steep slope of the Towans from 240m to 340m due to the irregularity of the ground and the scrub vegetation. Scanning with the 400MHz probe between 333m and 582m was possible over much of the transect but there are gaps due to dense or tall vegetation (Figure 46). The bedrock is shown for most of the scanned transect, but, as in the cores, there are no clear interfaces between the surface and the bedrock. There is irregularity in the solid surface indicating some hollows along the Towans summit. In particular, there is a depression at around 530m which is discussed below.

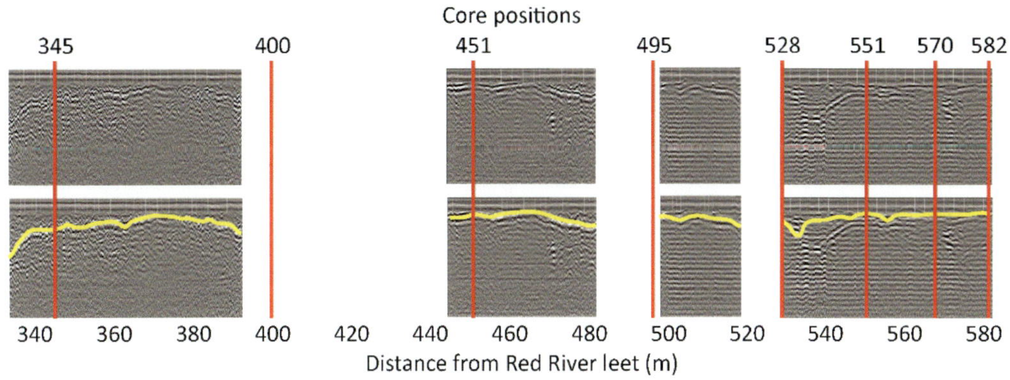

Figure 46. GPR between 333 and 582m (400MHz).
The gaps in the scans are where vegetation prevented access for scanning. The lower image shows the interface between the bedrock and overlying sandy sediments drawn onto the same scans.

The cores

Each core will be discussed individually with summary tables of laboratory investigations and mollusc diagrams. A zoning interpretation is given to the right of each summary and mollusc diagram which represents an analysis of all the analytical modalities used to provide a best fit to the stratigraphy. Numbers in the text in square brackets [ ] refer to these zones. When a sediment is named 'buried soil' this is determined on a combination of the mollusc numbers and sedimentology. For those cores which did not reach bedrock the lowest level reached has been designated zone [2]. The depths shown in all figures and tables pertaining to percussion cores relate to the depths corrected for compression and contaminating material, as described in Chapter III.

Some comments are necessary concerning the figures:

Summary diagrams

The first figure for each core shows a combination photograph of the core to demonstrate the stratigraphy; this is followed by bar charts of particle size and loss on ignition (LOI) (Figure 47) and line graphs of pH and the total number of molluscs at each level within the core, normalised to 500g samples. The depths are those below modern ground surface.

particle size						
clay	silt	very fine sand	fine sand	medium sand	coarse sand	gravel
0-2μm	2-63μm	63-125μm	125-250μm	250-500μm	500μm-2mm	>2mm

loss on ignition	
organic matter content	carbonate content

Figure 47. Keys for the particle size and loss on ignition charts

Mollusc diagrams

All the mollusc diagrams in the following two chapters show the absolute numbers of molluscs for each core, with percentage cumulative diagrams to the right of each figure. For every sample, the actual number of shells found and the sample weight is given for reference. Since the volume and weight of sediment in each sample varied considerably, some sediments being very thin (especially many of the organic silt levels) while others were thick, all the graphs show mollusc numbers normalised to 500g per sample. Dots on the diagrams indicate samples with ≤5 shells (normalised numbers). In many of the cores there were no molluscs in the deeper levels, so the diagrams only include samples down to the lowest level in which shells were found. The stratigraphic diagram included in each mollusc diagram uses the same colouring as in Figure 48. The Shannon and Brillouin diversity indices are shown graphically only when there are more than 10 shells in the relevant sample.

The habitat requirements on the diagrams are consistent in all the diagrams but are not intended to be prescriptive. It is accepted that the ecological niches of many molluscs are variable, and some of these variations are discussed in the accompanying text. References to specific habitats are only included in this Chapter when there are particular points to be made.

Chapter VI – Percussion cores

Percussion cores were obtained in two positions south of the Red River and in eight positions north of the river. All depths, both in the diagrams and the text, are those after correction for differential compaction of sand and organic silt.

Figure 48 shows the stratigraphy of the percussion cores. It can be seen that of the ten cores, the bedrock was only reached in six. Lines have been placed which link suggested equivalent sediment horizons.

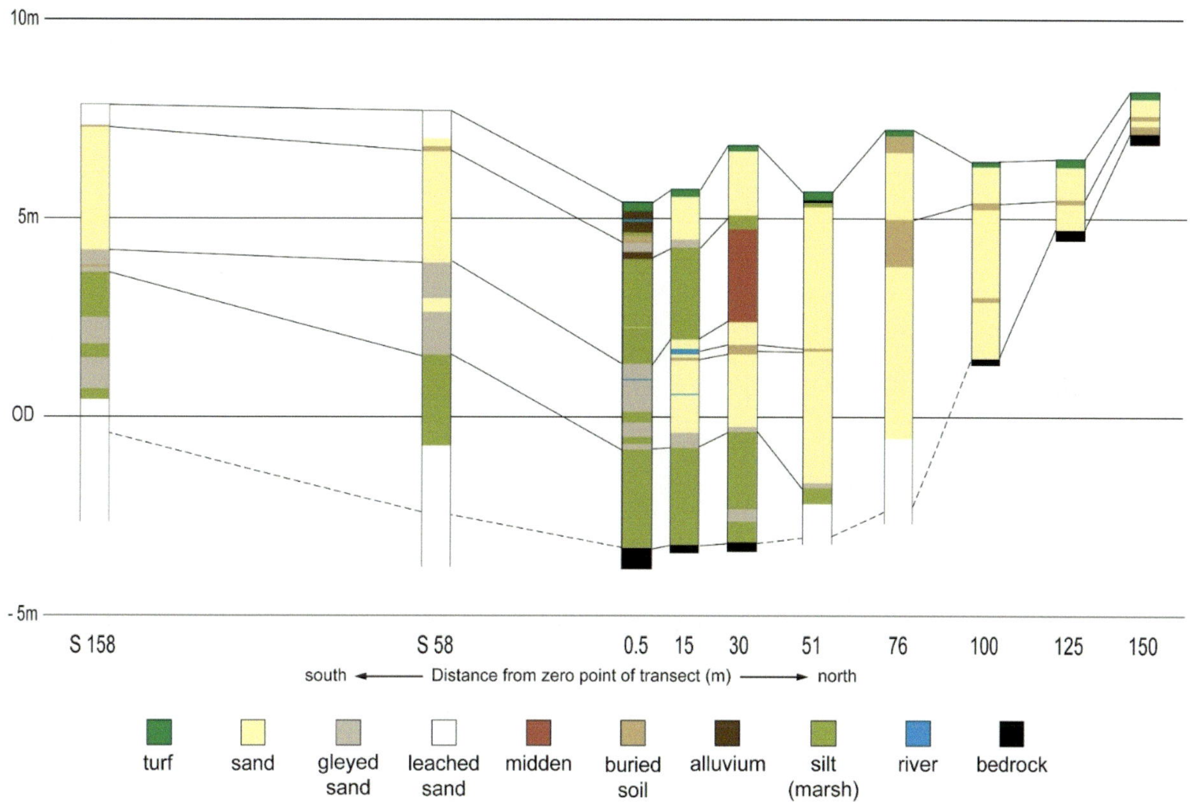

Figure 48. Stratigraphy of the percussion cores from Gwithian

South 158m core

Grid ref.: SW 58886.41876; modern ground level: 7.57m OD.
Mollusc numbers: Table 3; mollusc diagram: Figure 49; summary diagram: Figure 50.

Coring was extremely difficult deeper than about 500cm and abandoned at 714cm without reaching bedrock. Organic silt caused the sample tubes to vibrate sufficiently to prevent the percussion action of the Cobra unit from advancing the tube deeper than 485cm. The core tubes were replaced with decreasing diameters of gouges (5cm, 4cm) but even with these it proved impossible to core through the deeper silt levels and difficulty in extracting the corer precluded further attempts. The zone numbering therefore commences at [2] for the lowest level achieved, but it is not known whether sediment [2] lies directly on the bedrock or there are intervening deposits.

Table 3. Mollusc numbers for the South 158m core.

Depth in core (cm)	0-50	50-55	55-100	100-150	155-200	210-225	255-300	310-357	357-390	390-395	395-411	411-420	420-421	421-457	457-467	467-474	474-479	479-529	529-570	570-580	580-588	588-591	610-687	687-700
Corrected depth (cm)	**0-50**	**50-55**	**55-100**	**100-150**	**150-200**	**200-250**	**250-300**	**300-352**	**352-388**	**388-394**	**394-406**	**406-416**	**416-417**	**425-471**	**471-491**	**491-505**	**505-515**	**522-528**	**528-579**	**579-590**	**590-606**	**606-612**	**656-688**	**688-714**
Sample weight (g)	1137	111	1087	1184	1041	1242	1165	1303	791	118	321	225	14	553	176	115	101	371	681	290	162	90	1072	146
Munsell number	10YR/ 3/2		10YR 7/4						gley2 5/5B	10YR 3.2	gley2 5/5B	10YR 4/2	10YR 6/3	10YR 2/2	10YR 3/2	10YR 4/3	10YR 3/2	10YR 5/2	gley2 5/5B	10B 7/1	10YR 3/3	10YR 5/4	gley2 5/5B	10YR 3/2
Zone	**14**	**13**	**12**						**11**	**10**	**9**	**8**	**7**	**6**				**5**			**4**		**3**	**2**
Sediment	leached sand	buried soil	sand						gleyed sand	buried soil	gleyed sand	marsh	sand	marsh				gleyed sand			marsh		gleyed sand	marsh
Carychium minimum												3			32	14	3	2			5			
Galba truncatula										1					22	31	4				1			
Succineidae	2								3	2		2		2	16	63	38	4				6		
Cochlicopa lubrica		3					1								15	25	12	2			1			
Vertigo anitvertigo										1					6	4		2			4			
Vertigo pygmaea							1					2			10	4	4					1		
Pupilla muscorum							4	1				3												
Acanthinula aculeata															1									
Vallonia excentrica		36				1		1						1	4	2	5	2						
Punctum pygmaeum												1			3									
Discus rotundatus														2	17									
Vitrina pellucida					1																			
Vitrea contracta															5									
Aegopinella pura																10								
Nesovitrea hammonis												7		1	13	4		2			11			
Oxychilus cellarius						1																		
Milacidae														1										
Limacidae															1									
Cochlicella acuta	1	55	5	5	1		9	3	1			1												
Ashfordia granulata												4		1	62	1					2			
Candidula intersecta			2																					
Cernuella virgata		7	6						1					1										
Helicella itala				1	2	2	7																	
Trochulus hispidus	2									1														
Cepaea sp.		1				2									1									
Cornu aspersum			1																					
Pisidium personatum															13									
TOTAL SHELLS	**5**	**102**	**14**	**6**	**4**	**6**	**22**	**5**	**5**	**5**	**0**	**23**	**0**	**9**	**221**	**158**	**66**	**14**	**0**	**0**	**24**	**7**	**0**	**0**
Shells per 500g	2	459	6	3	2	2	9	2	3	21	0	51		8	628	687	327	19			74	39		
TOTAL SPECIES	**3**	**5**	**4**	**2**	**3**	**4**	**5**	**3**	**3**	**4**		**8**		**7**	**16**	**10**	**6**	**6**			**6**	**2**		
Seeds									+			+			+						+			
Wood																					+	+		+

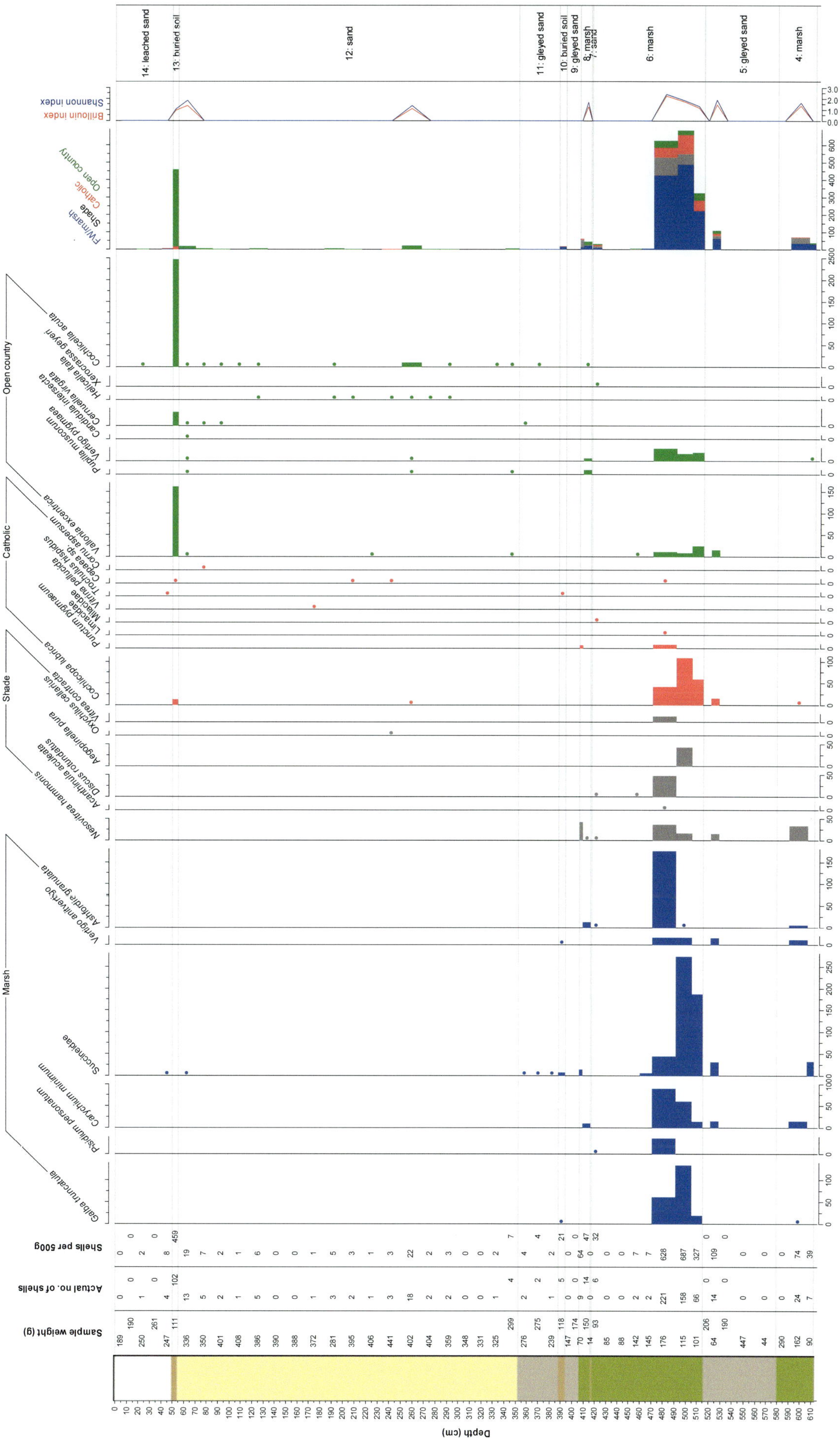

Figure 49. Mollusc diagram for the South 158m core. No molluscs were found below 612cm.

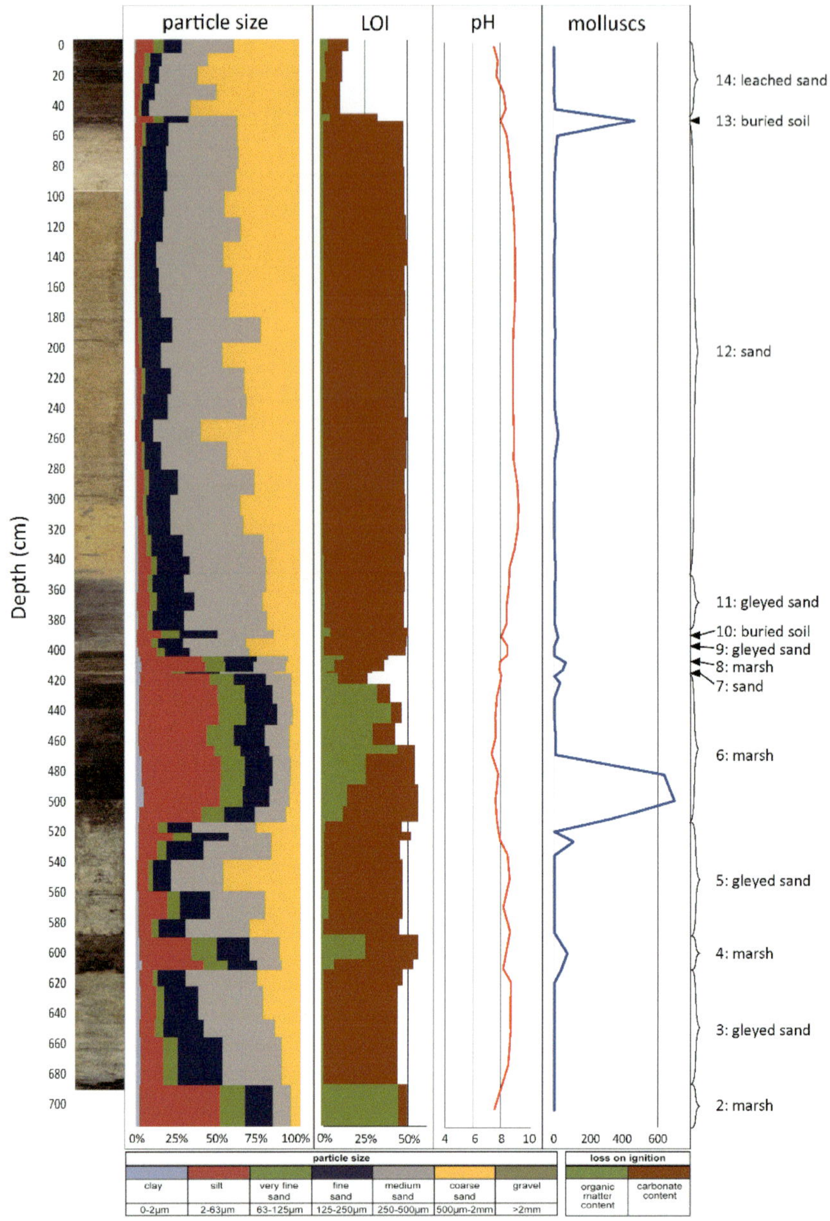

Figure 50. Summary diagrams for the South 158m core.

The most southerly core was obtained on the slope of the south side of the Red River valley (Figure 51) in a large field, now part of Churchfield Farm, and at the time of visiting grazed by cattle although with some root vegetable dumps. The field is generally level but slopes up to the south at the margin of the valley. Across the field is a bank of piled sand, about 100m in length although only 10m in width. This sand was bulldozed off the flat field surface in about 2000 by the farmer to create a windbreak for cattle on the field (A. James, pers. com.). There were no shells in the turf sample; there is no clear reason for this total absence of molluscan life, but it may be consequent to agricultural fertilisation, weed control and/or flooding with possible contamination from mining pollution. Extensive flooding was present on this field during early 2014 due to a very high water table after several weeks of severe rain (personal observation) and if this was a regular occurrence then conditions may be adverse for molluscs to live.

Figure 51. Coring at the South 158m location in the field south of the Red River.
The sand bank can be seen in the middle distance. The South 58m location is just beyond the sand bank.

The lowest depth reached was 714cm. The basal layer was organic silt [2] with 45% organic matter. This is overlain by gleyed sand [3] and then another horizon of organic silt [4]. These silt layers are interpreted as marsh/alder fen, the lowest level [2] containing large fragments of alder wood. Wood fragments were also present in layer [4], together with bog bean (*Menyanthus trifoliata*) seeds and it is in this layer that molluscs are first found, nearly all species associated with wet ground (*Carychium minimum*, *Galba truncatula*, Succineidae, *Vertigo antivertigo* and *Ashfordia granulata*), although a few dry ground species do occur (*Cochlicopa lubrica*, *Nesovitrea hammonis* and *Vertigo pygmaea*).

There is then a further deposit of gleyed sand [5], almost devoid of shells, before a thick marsh layer [6], the mid-portion of which contains large numbers of shells with wet ground taxa strongly predominating. The presence of large numbers of costataeidae and *Ashfordia* suggests that the vegetation in this area was tall and rank and that it was not regularly under water. Reeds and sedge are plentiful, and there is alder growing in the close vicinity, fragments of alder wood being present in many samples. Of interest here is the presence of *Pisidium personatum*, a bivalve mollusc associated with very poor habitats such as stagnant ponds, ditches and other wet ground habitats subject to regular drying. There are moderate numbers of catholic species but few open country shells in marsh horizon [6] and it may well be that these are allochthonous. Layer [7] is a very thin (1cm) sand horizon which may represent a storm event with blown sand covering the marsh; it is devoid of shells. The marsh, however, persists [8] for a short time before it is covered with a 12cm layer of gleyed sand [9], also without molluscs, and probably indicating the level of the water table at some time after deposition. This gives way to more stable conditions with a thin buried soil [10] at 388–394cm containing a few wet ground molluscs (*Galba truncatula*, Succineidae, *Vertigo antivertigo*), but further flooding returns with more gleyed sand [11].

Dryer conditions then become established, with a thick (302cm) layer of yellow sand [12]. The chronology of the interface between the gleyed sand [11] and the non-gleyed sand [12] can be inferred from the dating performed on the 30m core north of the river (see Chapter IX). In that core the highest deposit gleyed sand is dated to about 1300 cal BC, and it is therefore probable that the yellow sand [12] commenced accumulating in the middle Bronze Age. The consistency of this sand throughout its depth, and the noticeable paucity of molluscs suggests prolonged instability, and it is likely that there has been regular deflation and deposition over a long period, probably up to post-medieval times, and this thick horizon may be the result of one or many episodes of deposition spread over many centuries.

A thin buried soil [13] at 50–55cm indicates a period of greater stability; it contains open country molluscs, mainly *Vallonia excentrica* and *Cochlicella acuta*, the high numbers *Vallonia* suggesting grazed grassland. The topmost 50cm [14] of sand is almost devoid of molluscs and has a very low carbonate content; it is very similar to the 4–5cm thick turf block.

The overall interpretation of this location is that there are deep layers of marshland and alder fen [2, 4, 6] which were subject to periods of drying. There is no indication that the river course ever reached this far south, and regular overbank alluvium is unlikely due to the lack of true freshwater species. There were episodes of blown sand deposition which, at times, are gleyed implying high water table perhaps due to rising sea levels [5, 9, 11], but with at least one sufficiently stable episode for a shallow buried soil to accumulate [10]. This was followed by massive sand accumulation [12] without standing water and the relative lack of shells indicates that there were no significant periods of stabilisation until more recent times when there was a phase of stability with grazed grassland [13] and the land may have been used as pasture for farm animals. Finally, modern practices have created a largely sterile leached sediment [14] with little faunal activity.

South 58m core

Grid ref.: SW 58885.41976; modern ground level: 7.50m OD.
Mollusc figures: Table 4; mollusc diagram: Figure 52; summary diagram: Figure 53.

Table 4. Mollusc numbers for the South 58m core.

Depth in core (cm)	0-68	68-88	88-95	95-100	110-151	151-238	238-300	305-380	380-420	420-457	457-477	477-558	558-600	608-618	618-628	628-638	638-715
Correctd depth (cm)	**0-68**	**68-88**	**88-95**	**95-100**	**110-146**	**146-234**	**234-291**	**281-374**	**374-421**	**421-462**	**462-495**	**495-542**	**542-600**	**600-620**	**620-640**	**640-660**	**660-824**
Sample weight (g)	1923	422	198	118	950	1813	1504	1979	777	1004	827	1222	1163	157	134	125	1187
Munsell number	10YR 5/3	10YR 6/4		10YR 5/3	10YR 6/4				10YR 5/2	10YR 6/3	10YR 7/4	gley2 5/5B		7.5YR 4/2	7.5YR 4/3		
Zone	**9**	**8**	**7**		**6**				**5**		**4**	**3**		**2**			
Sediment	leached sand	sand	buried soil		sand				gleyed sand		sand	gleyed sand		marsh			
Carychium minimum														12	3	27	
Galba truncatula														51	3	3	
Succineidae										1		1	2	10			
Cochlicopa lubrica									1					4	1	2	
Cochlicopa sp.		1															
Vertigo antivertigo														4	1	2	
Pupilla muscorum							1										
Vallonia excentrica							3	1	1								
Punctum pygmaeum																1	
Vitrina crysatallina															1	4	
Nesovitrea hammonis															6		
Oxychilus cellarius																2	
Cochlicella acuta	8	9	12	9	5		5	1	2								
Ashfordia granulata														3	5	16	
Candidula intersecta	1																
Cernuella virgata	2	1	3	10	1		2		5	1							
Helicella itala	1						1				1						
Pisidium personatum														11	3		
TOTAL SHELLS	**12**	**11**	**15**	**19**	**6**	**0**	**12**	**2**	**9**	**2**	**1**	**1**	2	**95**	**23**	**57**	**0**
Shells per 500g	3	13	38	81	3	0	4	1	6	1	1	0	1	303	86	228	
TOTAL SPECIES	**4**	**3**	**2**	**2**	**2**	**0**	**5**	**2**	**4**	**2**	**1**	**1**	1	**7**	**8**	**8**	
Reeds/sedge															+	+	+
Seeds																+	
Wood																	+

This core was obtained from an area of level meadow between the Red River and the bank of sand mentioned above. Coring again failed to reach bedrock despite the use of gouges below 600cm, the corer being unable to penetrate deeper than 824cm.

The lowest recovered sediment [2] is organic silt, the lower half being acidic (down to pH 4.3) but becoming alkaline in its upper half (pH 8). Molluscs are only present in the non-acidic upper 60cm of this lowest unit, all being wet ground or catholic species with no taxa associated with open country. *Galba truncatula*, an amphibious mollusc of wet grasslands and poor freshwater habitats, is strongly dominant in the uppermost 20cm, with *Carychium minimum* and *Ashfordia granulata* a little deeper, perhaps indicating a slightly dryer habitat, although still with regular flooding. Reeds and sedge are present throughout this deposit with fragments of alder wood at 724–744cm.

Unlike the South 158m location there is no evidence of any marsh above 600cm depth. However, a thick layer of gleyed sand [3] overlying the marsh implies the position of the water table, perhaps long after deposition. This is covered by a non-gleyed horizon [4] indicating a dryer interval, but overlain by another

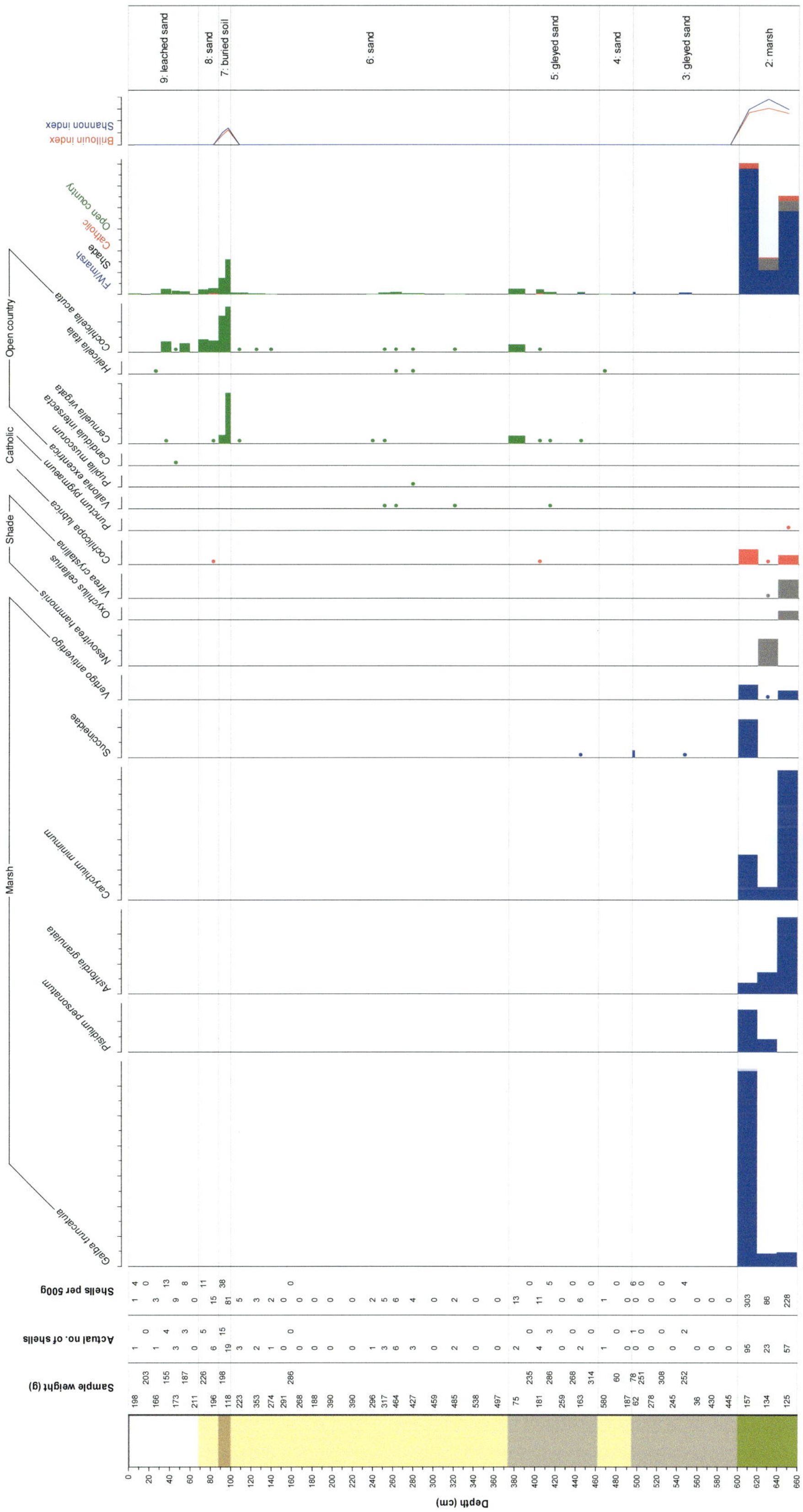

Figure 52. Mollusc diagram for the South 58m core. No molluscs were found below 660cm.

gleyed sand accumulation [5]. Above this there is a thick layer (254cm) of non-gleyed blown sand [6], corresponding to the equivalent accumulation at the South 158m position. None of these sands [3–6] contain more than very occasional molluscs, implying unstable ground with rapid accumulation of sand. It was proposed for the South 159m core that the corresponding sand commenced deposition around the middle Bronze Age but it is not possible to determine whether this thick accumulation is the result of one or more deposition episodes. A thin buried soil [7] is indicated by a slight increase in silt levels with open country molluscs (*Cochlicella acuta, Cernuella virgata*), although the absence of *Vallonia excentrica* makes grazing unlikely. Finally, there is a blown sand horizon [8] covered by the leached sand [9] similar to that found further south.

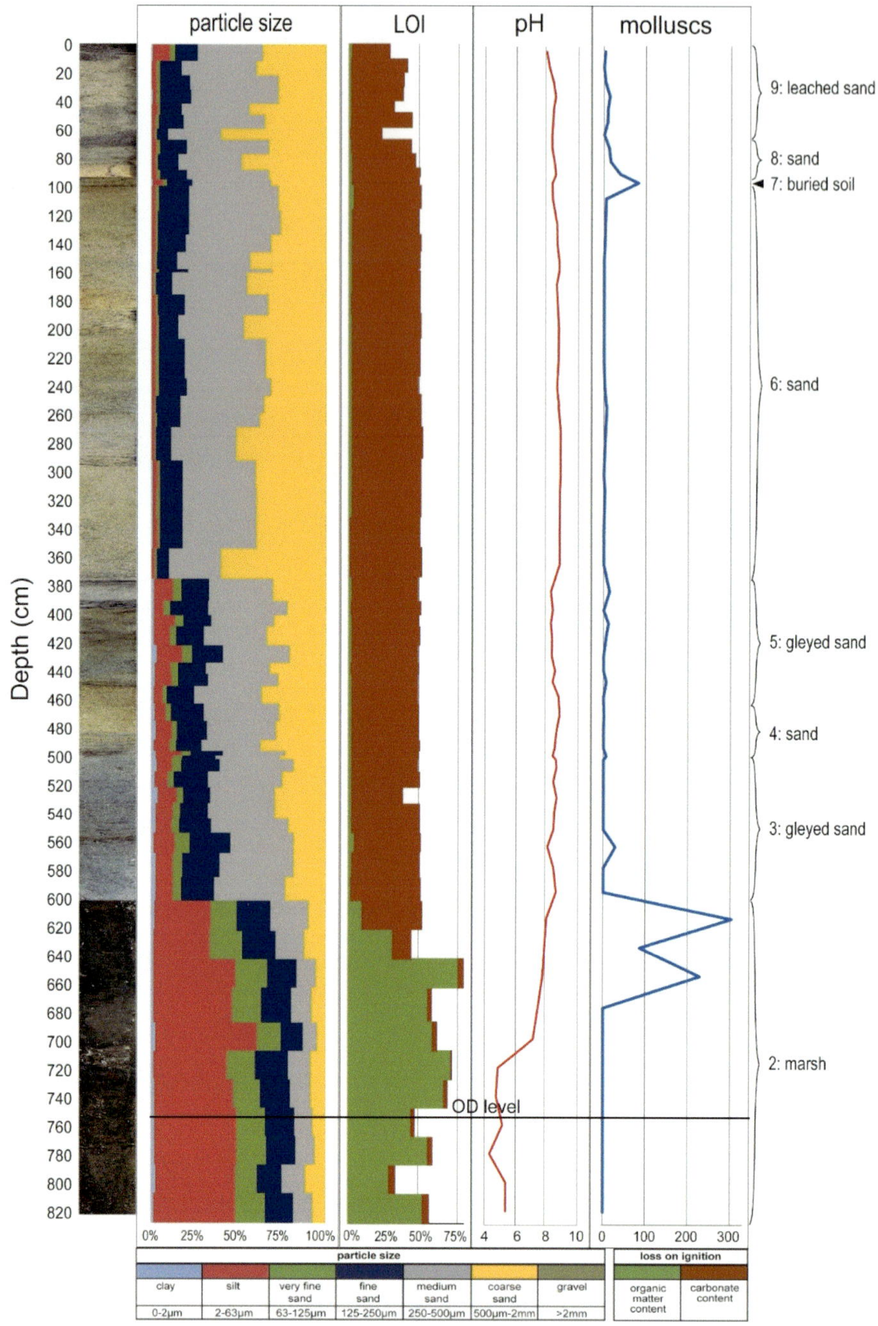

Figure 53. Summary diagrams for the South 58m core.

As at South 158m there is no suggestion of flowing river at this location. None of the sediments at either site contain coarse grained material which is found at other positions north of the present river, and at no level is there any mollusc associated with moving water. In addition, there is no suggestion that this area of the valley was ever estuarine; there are no clay layers similar to those described by Stephens (1899). It is possible that some clay could be found beneath the lowest levels reached by these new cores.

0.5M CORE

Grid ref.: SW 58886.42035); modern ground level: 5.42m OD.
Mollusc numbers: Table 5; mollusc diagram: Figure 54; summary diagram: Figure 55.

Table 5. Mollusc numbers for the 0.5m core.

Depth in core (cm)	0-21	21-41	41-46	46-74	74-83	83-92	92-100	115-135	135-142	142-195	195-200	254-262	262-267	267-268	268-279	279-358	358-376	376-380	380-390	390-393	393-486	486-489	489-500	537-570	570-580	580-587	587-850	850-900
Corrected depth (cm)	**0-21**	**21-41**	**41-46**	**46-74**	**74-83**	**83-92**	**92-100**	**100-122**	**122-139**	**139-264**	**264-275**	**275-294**	**294-306**	**306-307**	**307-331**	**331-400**	**400-419**	**419-424**	**424-435**	**435-438**	**438-515**	**515-522**	**522-541**	**541-577**	**577-594**	**594-607**	**607-849**	**849-900**
Sample weight (g)	129	262	56	665	223	189	214	397	132	383	56	92	83	21	144	384	388	97	245	64	1574	39	149	819	198	140	1649	1822
Munsell number	7.5YR 3/2	7.5YR 3/3	7.5YR 4/4	7.5YR 4.3	10YR 7/3			gley2 5/5B	10YR 3/1	10YR 3/1				10YR 6/2	10YR 3/1	10YR 2/1	10YR 6/2			10YR 5/2	gley2 5/5B	10YR 3/1	2.5Y 5/1	2.5YR 5/1	10YR 3/2	5YR 5/1	10YR 2/1	
Zone	**20**	**19**	**18**	**17**	**16**	**15**		**14**	**13**	**12**				**11**	**10**		**9**			**8**	**7**	**6**		**5**	**4**	**3**	**2**	**1**
Sediment	topsoil	alluvium	river	alluvium	marsh	buried soil		gleyed sand	alluvium	marsh				sand	marsh		gleyed sand			river	gleyed sand	marsh		gleyed sand	marsh	gleyed sand	marsh	killas
Carychium minimum										1	3	27	22		3						2	1						
Galba truncatula				1	8	3	4	3	1			23	37		21		1					19	1		1			
Radix balthica					1																							
Anisus leucostoma									2																			
Carychium sp.									2			8	5		7													
Succineidae				3	12	1	3		8		3	14	23		1													
Cochlicopa lubrica												7						1										
Cochlicopa sp.						1			4			4	2															
Vertigo anitvertigo												6	12								1		1		1			
Vertigo pygmaea							3											1										
Pupilla muscorum						1	3	1									1	4										
Vallonia excentrica						3	4	1									2	2			2	6						
Puntum pygmaeum																					1							
Discus rotundatus																									1			
Aegopinella pura																						1						
Oxychilus cellarius									1																			
Zonitoides nitidus									3			3	4												1			
Milacidae													1															
Euconulus alderi									1			7	3															
Cochlicella acuta					1	11	21	1	1								1	16	4		7	3			1			
Ashfordia granulata							1														1	1			2			
Cernuella virgata						1																						
Helicella itala				1		6	9	2	1								1	10	3									
Xderocograssa geyeri																	2	1	3									
Trochulus sp.																		1	1									
Cornu aspersum									1								1											
Pisidium sp.									8		4	55	40		11													
TOTAL SHELLS	**0**	**0**	**0**	**5**	**22**	**27**	**48**	**8**	**33**	**1**	**10**	**154**	**149**	**0**	**43**	**0**	**9**	**36**	**11**	**0**	**14**	**31**	**2**	**0**	**7**	**0**	**0**	**0**
Shells per 500 g				4	49	71	112	10	125	1	89	837	898	0	149		12	186	22		4	397	7		18			
TOTAL SPECIES				**3**	**4**	**8**	**8**	**5**	**12**	**1**	**3**	**8**	**9**	**0**	**4**		**7**	**8**	**4**		**6**	**6**	**2**		**6**			
Mytilus sp.																				+								
Reeds, sedge											+	+	+		+	+											+	
Seeds									+				+		+													
Wood																						+		+	+		+	
Bone																					+							

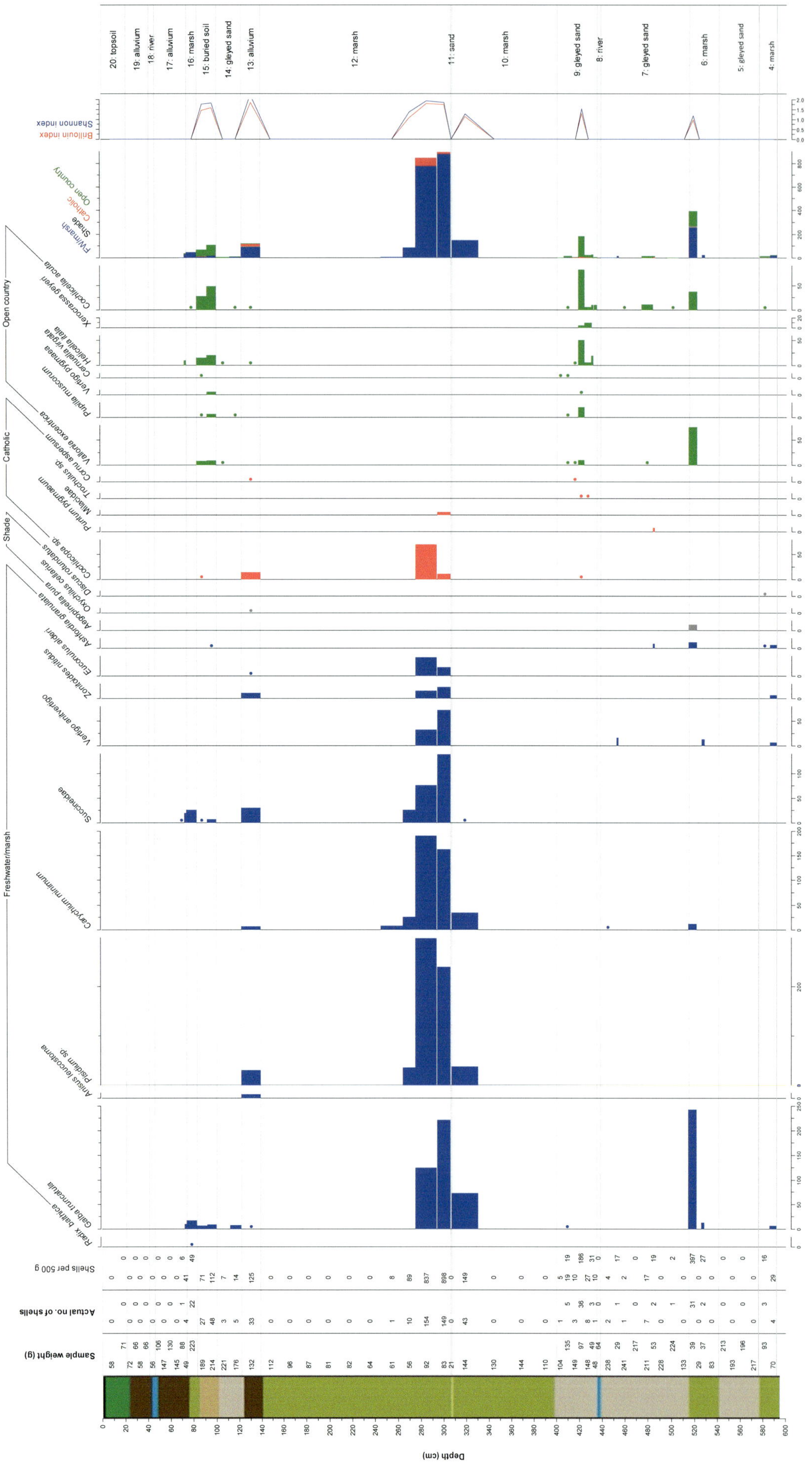

Figure 54. Mollusc diagram for the 0.5m core

No molluscs were found below 584cm (the diagram does not show the molluscs below 540cm).

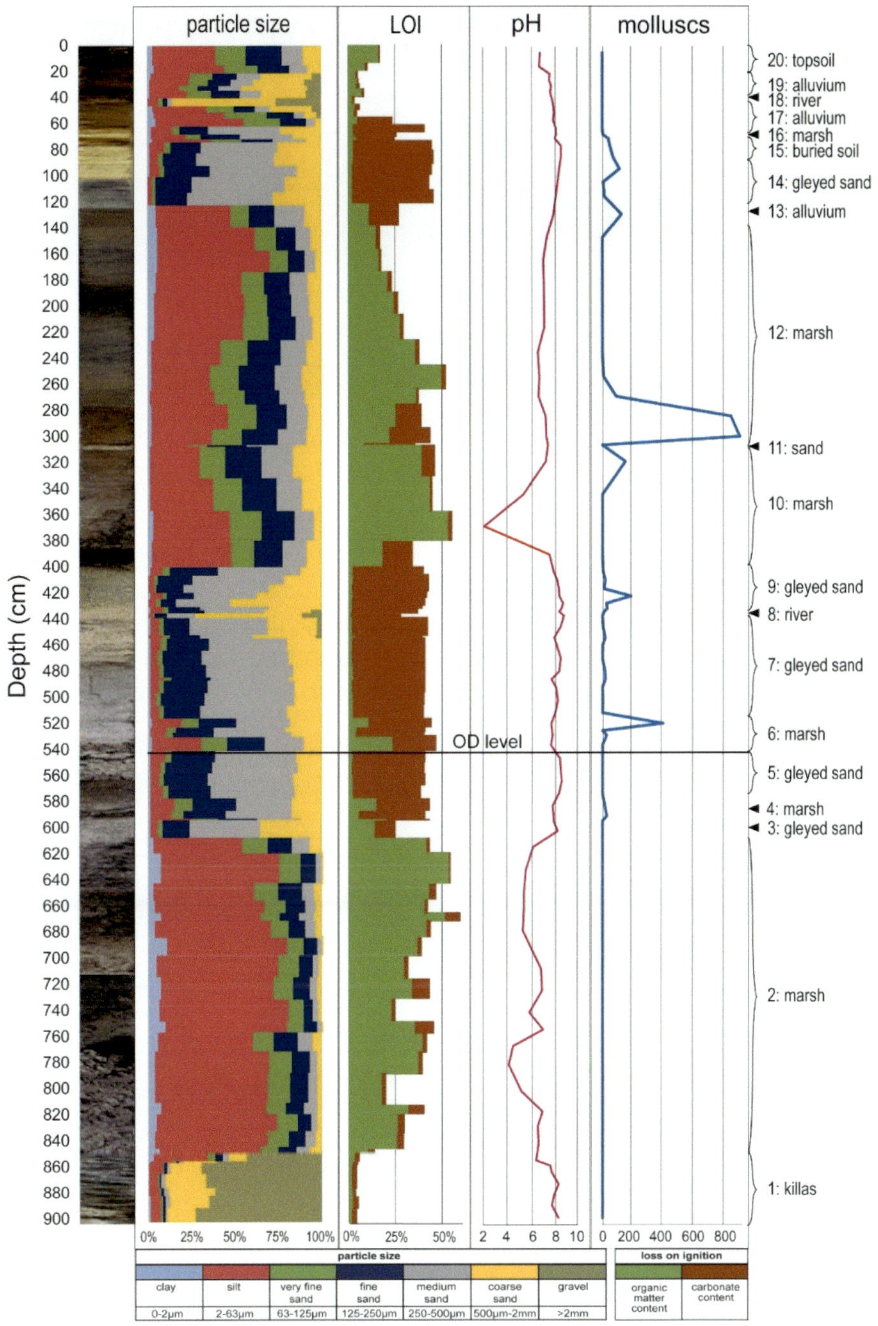

Figure 55. Summary diagrams for the 0.5m core.

This core was obtained as close as possible to the nineteenth century leet cut parallel to and north of the Red River channel. The ground between the leet and the river has been built up in modern times with a gravel track between the river and the leet, and no attempt was made to core close to the position of the present river. A wire fence is present on the north bank of the leet which was designated the '0m' position for the coring transect. At the present time there is tall lush vegetation in the meadow bordering the leet (Figure 56) including hemp agrimony (*Eupatorium cannabinum*), bird's foot trefoil (*Lotus corniculatus*), hemlock water-dropwort (*Oenanthe crocata*), common fleabane (*Pulicaria dysenterica*), sea bindweed (*Calystegia soldanella*), dead nettle (*Lamium* sp.), sea club-rush (*Scirpus maritimus*), creeping cinquefoil (*Potentilla reptans*) and silverweed (*Potentilla anserina*). There were only four molluscs in the turf block (*Aegopinella pura, Cernuella virgata* and *Cepaea nemoralis*) which is difficult to explain as the sediments are not acidic. Overbank flooding after heavy rain is known to occur (as happened in Spring 2013) but that alone should not account for the molluscan sterility.

Figure 56. Lush, dense, vegetation surrounding the 0.5m core position.

The core was driven 0.5m north of the fence, a distance which allowed working space for the coring equipment, and was successful to the bedrock at a depth of 849cm. This is a very complex core, with numerous stratigraphic levels. The bedrock [1] consists of the local killas; it is largely slate and represents Pleistocene Head. The lowest layer of deposited sediment [2], from 607–849cm, is organic silt. No molluscs have survived, which is not surprising as the pH is acidic throughout, dropping to pH 4.0 at one point. This sediment contains reeds and sedge, with tree fragments in several samples, with a tree branch/root filling the core tube at 745–754cm, identified as alder (*Alnus glutinosa*). As in the cores south of the river this is entirely consistent with marsh and alder fen.

A thin layer [3] (594–607cm) has slightly different sedimentary characteristics to adjacent deposits, with a higher mean particle size (Figure 57), and a higher proportion of coarse sand grains. The difference is not great and is considered to be the initial deposit of sand at this location.

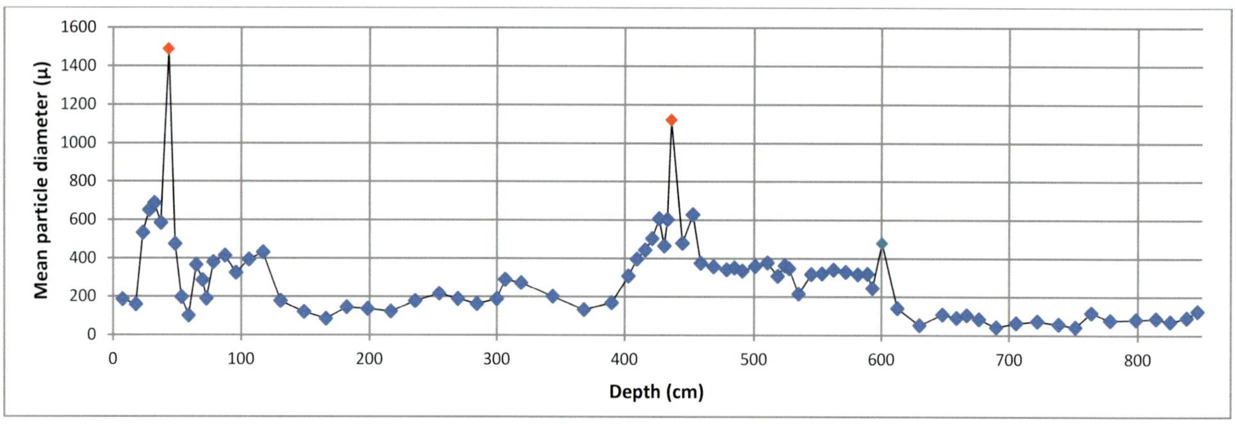

Figure 57. The mean particle size for each sample in the 0.5m core.
The points marked in red correspond to the assumed river sediments at 41–46cm and 435–438cm.

This is covered by another thin marsh layer [4] containing wood fragments. A very few molluscs (seven shells) are present in this horizon, mainly of marsh species. Sand [5] again accumulates at this location with alder fragments in its lower part; gleying indicates that it was regularly flooded with standing water or that the water table had risen to this level. The sand is overlain by a further probable marsh layer [6] with molluscs, mainly *Galba truncatula*, but with some *Vallonia excentrica* and *Cochlicella acuta*. The former implies continuing flooding but with dry periods allowing open country species to survive, although the latter may be allochthonous.

Further gleyed sand [7] with few shells covers the marsh. There is then another thin deposit [8] (435–438cm) with markedly coarser sand and gravel particles (Figure 57) which is likely to be riverine in origin. After a period of time the river changes course and there is a layer of gleyed sand [9], the mid-portion of which contains open country molluscs (mainly *Pupilla muscorum*, *Cochlicella acuta* and *Helicella itala*) implying a period of stability.

Wet conditions recur [10] with organic silt which is highly acidic in its lower part (pH as low as 1.9; the measurement was repeated on two samples to check this unexpectedly low reading) and it is not surprising that no shells have survived in this acidic environment. No explanation for the extreme acidity is evident although it could be related to stagnant water; it did not last long as the upper part of this organic silt contains wet ground molluscs, mainly *Galba truncatula*, *Carychium minimum* and *Pisidium* sp. The absence of dry land species suggests that these shells are autochthonous.

A very thin (1cm) layer of sand without any molluscs [11] interrupts the marsh deposits, consistent with a brief episode of blown sand deposition. It is covered by 170cm of organic silt [12] with abundant reeds and sedge throughout. Large numbers of molluscs are present in the lowest 30cm, mainly wet ground species (*Carychium minimum*, *Galba truncatula*, Succineidae, *Vertigo antivertigo*, *Zonitoides nitidus* and *Euconulus alderi*) and with a few dry ground shells (*Cochlicopa* sp., Milacidae), the whole assemblage indicating standing water for much of the time, but with areas of land out of water where riverside species could thrive. The molluscs then disappear, although marshy conditions persist, implying that water levels have risen sufficiently that these amphibious molluscs can no longer survive. Conditions were unstable as no freshwater molluscs appear. Only in the uppermost 17cm of the silt [13] do wet ground molluscs return, but the reduced organic matter content suggests that this may be overbank alluvium rather than continuing marsh.

Gleyed sand [14] with few molluscs covers the alluvium but there is then a period of greater stability [15] with open country molluscs appearing (*Cochlicella acuta*, *Helicella itala*, *Vallonia excentrica*, *Vertigo pygmaea*) indicating that this is an interlude of drier ground although still with moderate wetness. Marshy conditions return [16] with *Galba truncatula* and Succineidae; above this, molluscs disappear and there are no shells at all in the top 68cm of the sediments. It is likely that some of these superficial silty deposits are alluvial [17, 19] but a river deposit [18] at 41–46cm (Figure 57) indicates that for a period the river course moved to this site and was sufficiently close for regular overbank alluvium to cover the surrounding valley both earlier and later in time. The top 15cm [20] has a higher organic matter content than deeper levels and with very low carbonate content. No pollen has survived in a sample at 10–11cm depth, which together with the lack of molluscs further supports some local alteration to the conditions in these superficial deposits, and it is conjectured that this may be related to mining activity or regular variation in water table level.

In summary, this core, very close to the modern river course, shows that this north side of the valley was very marshy for much of the Holocene, but with intermittent periods of drying with layers of blown sand. The river itself has been at this location at least twice. The short period when the sediments are highly acidic [10] is difficult to explain.

Pollen analysis was performed on 20 samples from this core by Dr C. R. Batchelor (University of Reading) and interpreted on a time/depth scale. The findings are not included in this section but discussed in detail in Chapters VIII and IX when discussing the chronology of the valley.

15M CORE

Grid ref.: SW 55886.42049; modern ground level: 5.55m OD.
Mollusc numbers: Table 6; mollusc diagram: Figure 58; summary diagram: Figure 59.

Table 6. Mollusc numbers for the 15m core.

Depth in core (cm)	0-19	19-77	77-100	100-124	124-141	141-145	145-155	155-162	162-185	185-196	196-200	240-248	248-252	252-256	256-265	265-292	292-350	350-355	355-369	369-375	375-458	458-461	461-500	522-524	524-566	566-600	630-880	880-900
Corrected depth (cm)	0-19	19-77	77-100	100-126	126-145	145-153	153-171	171-184	184-227	227-248	248-274	274-289	289-297	297-304	304-321	321-372	372-396	396-402	402-417	417-424	424-507	507-510	510-554	554-556	556-603	463	640-880	880-900
Sample weight (g)	108	949	422	582	334	38	74	42	70	40	49	39	51	21	45	108	511	152	331	138	1755	97	699	56	1148	#REF!	735	176
Munsel number	7.5YR 3/1	10YR 7/6			gley1 5/10G	7.5YR 2.5/1		10YR 2/1		10YR 2/2	7.5YR 2.5/2	5YR 3/1	7.5YR 4/2	5YR 2.5/1			10YR 6/2	10YR 5/6	10YR 7/4	10YR 6/4	10YR 4/3	10YR 5/6	10YR 7/4	10YR 6/6	2.5Y 7/3	gley1 5/10Y	7.5YR 3/2	
Zone	16	15			14	13											12	11	10	9	8	7	6	5	4	3	2	1
Sediment	topsoil	sand			gleyed sand	marsh											sand	river	sand	buried soil	sand	river	sand	river	sand	gleyed sand	marsh	killas
Potamopyrgus antipodarum				1																								
Carychium minimum							1	1		8	76	34	88	16	2													
Galba truncatula						4	10	1		3	55	47	56	31	1													
Anisus leucostoma						4	12	2					1		1													
Succineidae	9						5	1			10	12	24	4														
Cochlicopa lubrica												5	8							3								
Vertigo anitvertigo										2	5	3	13	1														
Vertigo pygmaea			1																	2								
Pupilla muscorum		3	1	2	10												1			11								
Vallonia excentrica		1			4					1			1					1		9								
Vitrina pellucida			1																									
Zonitoides nitidus						2																						
Limacidae													1	1														
Euconulus alderi													1	1														
Cochlicella acuta	9	12	12	1	27	4	4				1	2	2	1			2	1	4	24	5							
Ashfordia granulata													1	1							1							
Candidula intersecta	2																											
Cernuella virgata		3	2		2						2	3																
Helicella itala	4	7	16	1	19						4			2							4							
Trochulus hispidus													9															
Pisidium personatum						9	30	5		12	54	17	38	14	6													
TOTAL SHELLS	24	26	33	5	62	23	62	10	0	26	207	123	243	72	10	0	3	2	4	49	10	0	0	0	0	0	0	0
Shells per 500 g.	111	14	39	4	93	303	419	119		325	2112	1577	2382	1714	111		3	7	6	178	3							
TOTAL SPECIES	4	5	6	4	5	5	6	5		5	8	8	13	10	4		2	2	1	5	3							
Mytilus sp.																	+	+	+	+	+	+	+					
Patella sp.																			+									
Reeds/sedge							+	+	+	++					+	+	+										+	
Seeds							+					+																
Caddis fly cases							+				+					+												
Wood									+																		+	
Hazelnut																											+	

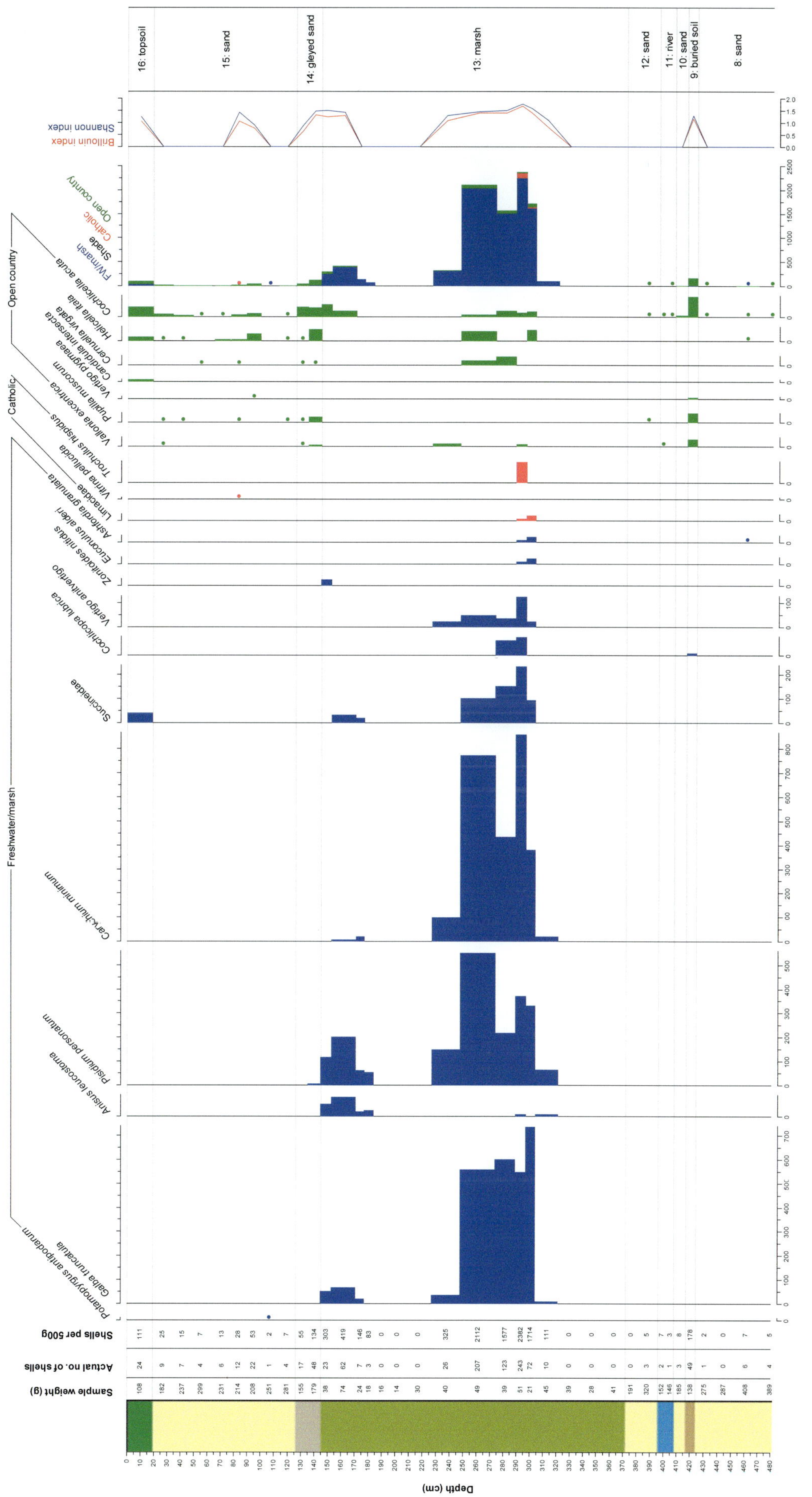

Figure 53. Mollusc diagram for the 15m core.
No molluscs were found below 483cm.

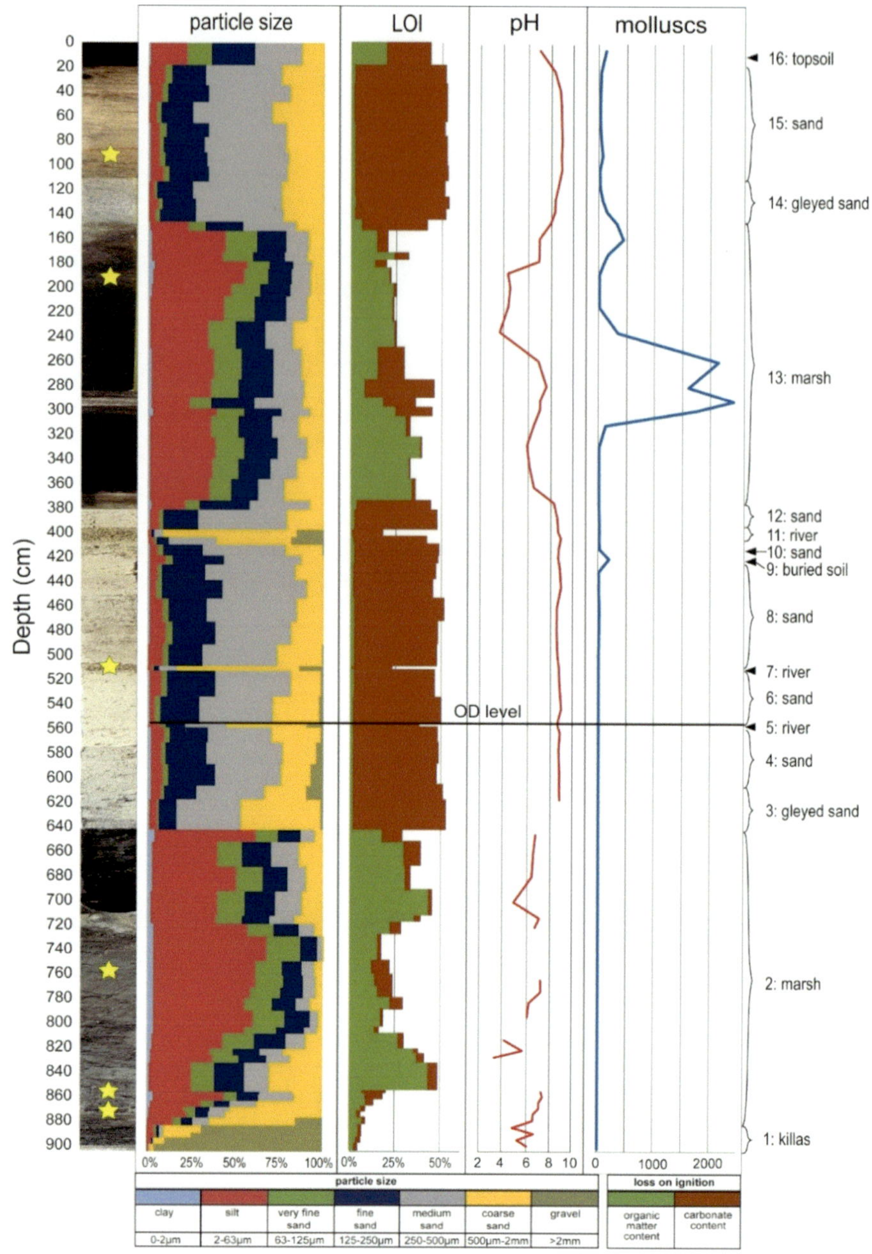

Figure 59. Summary diagrams for the 15m core.
☆ = diatom analysis. The gaps in the pH graph are where there was insufficient material for analysis.

This core location is at the junction of the tall riverbank vegetation and the shorter but rank grass of the meadow between the Red River and the B3301 road (Figure 60). Flowering plants include hemp agrimony (*Eupatorium cannabinum*), creeping cinquefoil (*Potentilla reptans*) and silverweed (*Potentilla anserina*).

Figure 60. The 15m core is marked by a white peg in the centre of the photograph. The long grass at the margin of the taller vegetation is clearly seen.

The core reached the killas at a depth of 880cm. Coring with the 5cm tube equipment was only possible down to 600cm and gouges were used for deeper sediments: 6cm to 700cm, 5cm to 740cm and 4cm to 895cm.

The bedrock [1] is killas, as found elsewhere. The basal deposit on the bedrock is a thick horizon (346cm) of organic silt [2], although the proportion of organic matter varies considerably at different levels, being low (<10%) in the deeper layers and increasing abruptly to nearly 40% at 852cm depth. No molluscs have survived in these silts which are mainly acidic. There are wood fragments at many levels, as well as reeds/sedges, and a hazelnut was noted at 869–874cm. The findings are entirely consistent with marsh/alder fen. The differing proportions of silt and organic matter indicate that there was regular change in the landscape, with periods when water may have been deeper than at other times, with variable amounts of relatively dry land.

There is abrupt change at 640cm when sand begins to be deposited. The lowest sands [3] are gleyed indicating the persistence of regular flooding or standing water but above this yellow sand [4] shows that the land has dried out. Organic matter almost disappears and carbonate content increases to 40%, although there are still no surviving molluscs, despite the pH being around 9 throughout.

The river channel [5] moves to this location, suggested, as in the 0.5m core, by increase in grain particle size at 554–556cm (Figure 61). The river then moves elsewhere and there is further lens of sand [6] before the river returns to this location [7] (507–510cm). Shells of open country species begin to appear in the next layer of sand [8], although numbers are small and they may be the result of bioturbation from a more stable sand horizon [9] which contains a moderate number of molluscs consistent with a buried soil, with increasing silt but no appreciable change in sediment colour. These shells are of open country species, mainly *Cochlicella acuta*, *Vertigo pygmaea*, *Pupilla muscorum* and *Vallonia excentrica* consistent with newly stable short grassland.

Fragments of mussel shells (*Mytilus* sp.) are present in many of the samples from 554cm [12] up to 372cm [6], around the time that the river was in this vicinity. Whether these were dropped by the inhabitants of the nearby settlement or deposited during flooding episodes cannot be determined; the former might seem more likely as no other marine shells were found (apart from one limpet fragment) and which might have been expected if this was a natural occurrence.

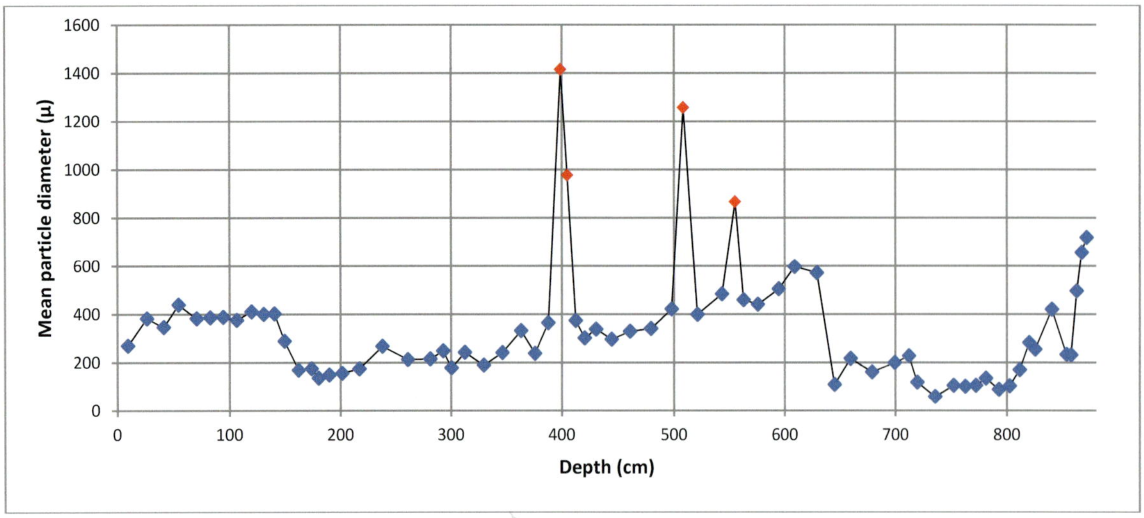

Figure 61. The mean particle size for each sample in the 15m core.
The points marked in red correspond to the assumed river sediments at 396–408cm (2 dots), 507–510cm and 554–556cm.

There is a further thin horizon of sand [10] before the river moves to this site [11] for the third time (396–408cm), again with high mean particle size (Figure 61). Sand [12] covers the river horizon but marsh conditions return [13], with another thick organic silt layer, 227cm in depth. There are no molluscs in the lowest 50cm of this marsh where the pH falls to 6.1, although reeds/sedge are present. In time, conditions change sufficiently for molluscs to thrive in very large numbers, up to 2382 per 500g. These are nearly all wet ground species (mainly *Carychium minimum*, *Galba truncatula*, Succineidae, *Vertigo antivertigo*, *Euconulus alderi*, *Ashfordia granulata* and *Pisidium personatum*) with only very few dry ground shells. The marshy areas with both standing water and some dryer areas found nearer the river clearly extend this far. The pH, having risen to 7.7 in the shell-rich sediments, falls to 4.4 in the higher marsh levels with corresponding absence of shells. The pH then rises again, with return of wet ground molluscs, although in smaller numbers than in the deeper levels. The presence of *Anisus leucostoma* is interesting, as this amphibious mollusc prefers sites with summer drying, strongly implying that the marsh conditions are variable at different times of the year.

Sand covers the organic silt but the lower levels are gleyed [14] although containing xerophilic molluscs, especially *Cochlicella acuta* and *Helicella itala*, indicating some degree of stabilisation. The overlying yellow sand [15] generally contains few molluscs despite having a carbonate content of over 40%. Of relevance is the finding at 100–126m of a single specimen of *Potamopyrgus antipodarum*, a freshwater snail introduced into Britain in the mid-nineteenth century and probably washed here from the nearby river. It is noted that this is one of the common species in the modern leet water sample.

The top 19cm of this core [16] contains a few molluscs including Succineidae, *Cochlicella acuta*, and *Helicella itala* consistent with a habitat of long grassland adjacent to tall mixed vegetation. Two shells of *Candidula intersecta* are also present, a species absent from the older, deeper, levels.

This core north of the present river course shows a similar profile to the adjacent 0.5m core, with an initial marshy environment subject to regular flooding from the river. The early marsh is followed by somewhat dryer conditions, only to be succeeded by more marsh. The course of the river is variable, moving here on three occasions, but this seems to be its most northern limit. At no time has there been really stable dry land this close to the river which would have allowed terrestrial molluscs to thrive, and even in modern times there is a very marked absence of molluscs, only four specimens being present in the modern topsoil at 0.5m and three at 15m. Why this should be is unclear, as conditions seem suitable and are currently not acid. However, fluctuation in water table levels or pollution from mining contamination are possible explanations. The very marked acidity of some of the higher marsh sediments is likely to be caused by the onset of mining activity, and will be discussed in Chapter XIII

Diatom analysis

This was the only location at which samples were analysed for diatoms. The aim was to see if there was evidence of marine or freshwater diatoms which may have indicated that this area of the valley was estuarine at some time. Six slides prepared in Reading were kindly assessed and interpreted by Dr Nigel Cameron (University College London).

Three samples from the deep marsh layer [2], at 869–874cm, 855–860cm and 758–768cm, all contained similar diatom assemblages with freshwater (mostly acid water) taxa:

Eunotia (e.g. *E. pectinalis* v. *minor, E. pectinalis* v. *ventralis, E. incisa* and others)
Pinnularia spp.
Cymbella (*Reimeria*) *sinuata*
Frustulia sp.
Tabellaria quadiseptata (possibly *T. fenestra*)
Fragilaria exigua
Fragilaria virescens
Gomphonema parvulum
Achnanthes lanceolata
Cocconeis placentula
Meridion circulare
Diatoma sp.
Navicula perpusilla

At the river level [5], 507–510cm, there were very few valves or fragments but those present included:

Tabellaria flocculosa
Hantzschia amphioxys
Pinnularia cf. *irrotata*
other indeterminate fragments

Diatoms were absent in two more superficial samples: marsh [11], 248–274cm and sand [13], 89–100cm.

Dr Cameron states 'There is no evidence for marine incursion from these diatom assemblages, with no marine or marine-brackish taxa present. Freshwater taxa with higher salinity optima, e.g. *Rhopalodia gibberula, Synedra tabulata, Synedra pulchella* are present but rare and do not necessarily indicate any significant increase in salinity.'

The conclusions to be drawn from this limited diatom study support those found in the sediments and the mollusc findings - that there has never been marine incursion at the site of this core, only about 20m north of the present river course. It is, however, suggested that the marine inlet may have extended this far inland elsewhere in the valley (see discussion in Chapter XIV).

30M CORE

Grid ref.: SW 58886.42064; modern ground level: 6.71m OD.
Mollusc numbers: Table 7; mollusc diagram: Figure 62; summary diagram: Figure 63.

Table 7. Mollusc numbers for the 30m core.

Depth in core (cm)	0-15	15-100	108-171	171-175	175-188	188-253	253-320	320-363	363-396	387-456	456-460	460-467	467-481	481-537	537-564	564-665	665-676	676-862	862-889	889-997	977-1000
Corrected depth (cm)	**0-15**	**15-100**	**100-172**	**172-180**	**180-207**	**207-271**	**271-340**	**340-387**	**387-435**	**435-491**	**491-506**	**506-514**	**514-530**	**530-571**	**571-601**	**601-693**	**693-705**	**705-896**	**896-926**	**926-997**	**977-1000**
Sample weight (g)	117	1635	1264	64	157	1314	1456	1005	740	1286	82	171	375	940	694	2164	270	1526	452	396	685
Munsell number	10YR 3/2	10YR 7/4	10YR 6/3	10YR 5/3	10YR 3/2	10YR 5/2	10YR 6/2	10YR 6/3	10YR 6/2	10YR 7/2	10YR 4/3	10YR 5/2	10YR 7/1	10YR 7/4	10YR 6/3	10YR 7/4	gley2 5/5B	10YR 2/2	10YR 61/1	10YR 2/2	
Zone	**12**	**11**		**10**		**9**				**8**	**7**		**6**				**5**	**4**	**3**	**2**	**1**
Sediment	topsoil	sand		marsh		enriched sand (midden)				sand	buried soil		sand				gleyed sand	marsh	gleyed sand	marsh	killas
Carychium minimum				10	59	4	3	22			58	5									
Galba truncatula					24				3		4										
Radix balthica									6												
Anisus leucostoma					10			2													
Gyraulus crista					10				3												
Gyraulus laevis								1													
Succineidae				3	37	8			1												
Cochlicopa lubrica		3			23	11	3	2	1		25	14	2			1					
Cochlicopa lubricella							1														
Cochlicopa sp.	1	1		6	65	12	7	4	2		18	18	1								
Vertigo anitvertigo					16			1			11										
Vertigo pygmaea	1	1			18	2	1		4		4	7									
Vertigo sp.					26						3										
Pupilla muscorum	1	7	4		1	23	4	5	6	8	2	16	5		6						
Lauria cylindracea								1	51	11	156	59	10		10						
Vallonia excentrica	6	13	2		27	91	56	28													
Punctum pygmaeum	1	1				2		3	2		16	4	1		1						
Discus rotundatus							7														
Vitrina pellucida		1									1										
Vitrea crystallina							1														
Aegopinella nitidula						1	18	3		3	16	10		1	1						
Aegopinella pura							4				6										
Nesovitrea hammonis						2															
Oxychilus cellarius												2									
Zonitoides nitidus					9																
Milacidae											7										
Limacidae					1	1	1	2	2		1	1									
Euconulus alderi											2										
Cochlicella acuta	24	45	22	12	144	421	178	133	85	37	53	88	26		26	1					
Ashfordia granulata					10	22	52	62	7	5	24	20									
Candidula intersecta	2																				
Cernuella virgata	8	14	9			65		3													
Helicella itala	2	34	15	1		10	10	8	17	27	19										
Xerocrassa geyeri									2	7	3										
Trochulus sp.			1	8	79	23	346		1												
Cepaea sp.																					
Cornu aspersum						frags	frags	frags	2												
Pisidium personatum																					
TOTAL SHELLS	**46**	**120**	**53**	**40**	**559**	**698**	**693**	**282**	**200**	**98**	**429**	**244**	**45**	**1**	**44**	**2**	**0**	**0**	**0**	**0**	**0**
Shells per 500g	197	37	21	313	1780	266	238	140	135	38	2616	713	60	1	32	0					
TOTAL SPECIES	**9**	**10**	**6**	**6**	**15**	**16**	**16**	**15**	**16**	**7**	**18**	**11**	**5**	**1**	**5**	**1**					
Mytilus sp. (g)		+	+			8.6	24.2	29.5	9.7	5.5											
Patella sp.						+	+	+	+												
Nucella lapillus						+	+	+	+	+											
Barnacles								+	+												
Fish scales								+	+												
Seeds				+	+																
Wood																		+		+	
Reeds/ sedge																		+		+	
Animal bone							+	+	+	+											
Charcoal									+												

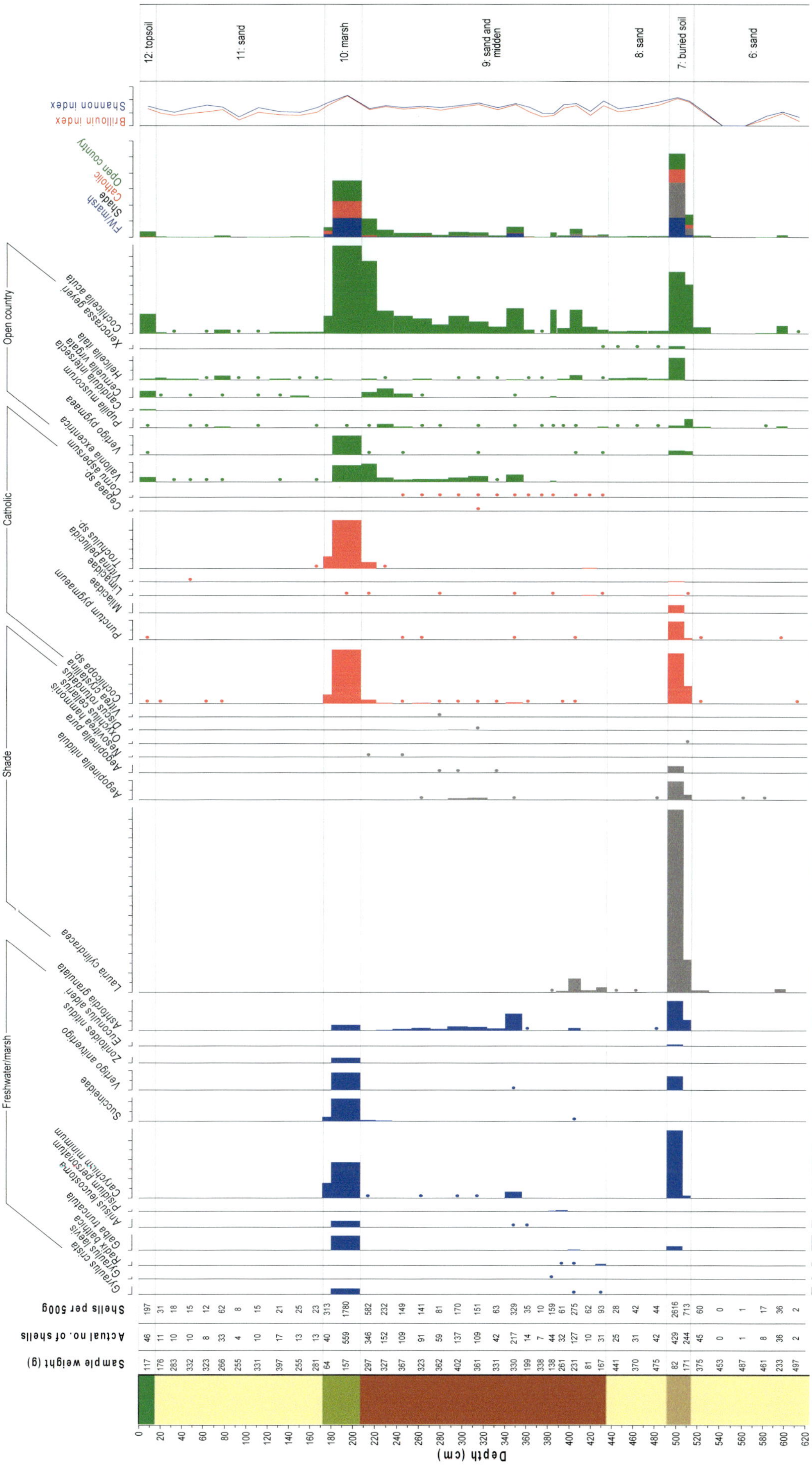

Figure 62. Mollusc diagram for the 30m core. There were no molluscs below 621cm.

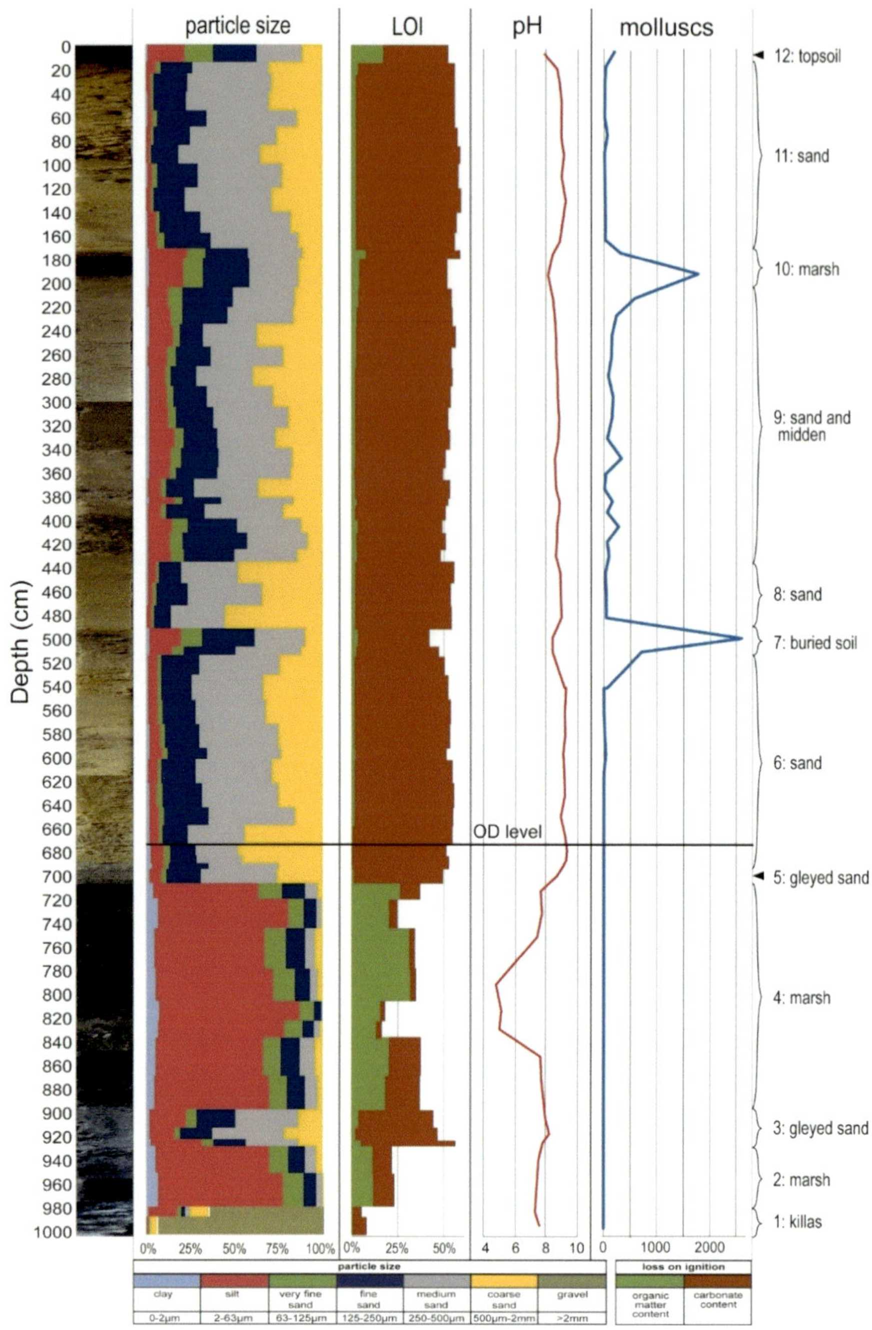

Figure 63. Summary diagrams for the 30m core.

The 30m core is the one on which OSL and radiocarbon dates have been obtained. The results are not included here but will be discussed in Chapter IX when the chronology of the valley is considered.

The core, taken from the meadow just south of the B3301 road, was from an area of tussocky grass growing to a height of about 30cm, and therefore providing moderate shade (Figure 64). Lady's bedstraw (*Galium verum*) and common knapweed (*Centaurea nigra*) were found flowering in the immediate vicinity. The turf contained a high number of shells (289 per 500g) and the highest number of taxa (17) of any site, as well as the highest species diversity (Shannon index 2.35); there is a good range of taxa with differing ecological preferences. The meadow here has not, in recent years, been regularly grazed or cultivated and it may be that the lack of disturbance has allowed the very rich molluscan fauna to develop. This assemblage is in distinct contrast to the absence of molluscs in the turf at the 15m location; the 30m site is on a slight elevation, with the ground level 1.16m higher than at the 15m position, and therefore less liable to flooding. This elevation may have allowed the immediate surroundings to be spared the effects of regular inundation and perhaps pollution from the Red River.

The percussion core reached killas [1] at a depth of 977cm. The basal sediment is 51cm of organic silt [2], likely to be marsh, and is overlain by 30cm of gleyed sand [3]. There is no equivalent sandy layer in the 0.5m or 15m cores and it is therefore likely that this position represents the margin of dryer land adjacent to the earliest river bank. The sand is overlain with a thick layer of organic silt [4] containing wood and reeds, more abundant in the upper than the lower portion, and is interpreted as marsh. The pH at the top and bottom is between 7 and 8, but falls to 4.8 centrally.

Figure 64. The 30m core site showing rank tussocky grass but no taller vegetation.

The marsh is covered with a 191cm deposit of blown sand, the basal 12cm of which is gleyed [5] indicating continued flooding, while the majority is yellow sand [6]. Molluscs have survived in the upper half of the sand, although are few in number indicating instability. Mollusc numbers increase at the top of this sand which is overlain by a buried soil [7], with an increase in proportion of silt. This deposit is very rich in molluscs, both in terms of numbers (up to 2616 per 500g) and of taxa (19 species) and with high diversity (Shannon index 2.17 in the upper half). The assemblage includes taxa from all habitat groups. *Carychium minimum*, *Galba truncatula* and *Vertigo antivertigo* imply wet ground, but this is probably intermittent. There are large numbers of open country and catholic species with a few, such as *Aegopinella nitidula* and *A. pura* requiring some shade which may be found at the base of rank vegetation. *Pupilla muscorum*, an early coloniser of rough ground, is present but *Vallonia excentrica* is absent, suggesting that the ground was not grazed. The remarkable mollusc in this deposit is *Lauria cylindracea*, which accounts for over a third of all shells present, in contrast to adjacent cores where this species was not found. Although classified in Figure 62 as a shade species, *Laura* is a mollusc characteristically associated with stone walls (Evans 1972, 151). The presence of field walls or possibly buildings close to the river which, at the time *Lauria* is abundant, was probably at the 0.5m location, seems a plausible explanation; they may have been constructed along the river bank, perhaps for stock control. At Brean Down in Somerset, a similar superabundance of *Lauria cylindracea* was found associated with a stone structure (Bell 1990, Fig. 156).

This stabilisation horizon is overlain by a blown sand layer [8] with relatively few molluscs, although still with a *Lauria* presence. Small numbers of the mollusc *Xerocrassa geyeri* are present in this sand, but in insufficient numbers to draw conclusions about whether they are autochthonous to this layer or intrusive from a lower deposit.

Above this is a thick (228cm) deposit of sand [9] but with different characteristics. There is slightly more silt than in the lower sand. *Lauria cylindracea* disappears and is replaced by *Vallonia excentrica* which is absent at all deeper levels, and with only a few *Pupilla muscorum* suggesting that there may have been short grazed grassland at this time. Totally different from the deposits above or below is the presence of numerous marine molluscs and other marine faunal fragments. Mussels (*Mytilus* sp.), limpets (*Patella* sp.) and dog whelks (*Nucella lapillus*) are present throughout. At one level (381–387cm) mussels comprise nearly 9% by weight of the whole sample. There are barnacle plates, fish bones and fish scales in some samples. The consideration of marine incursion over this period to account for the marine molluscs can be discounted as there are virtually no small marine shells; non-marine molluscs could not survive in a saline environment and it is very unlikely that they are all allochthonous. Animal bone fragments, not identifiable to species but of a size consistent with pig or sheep, are present at most levels with some wood charcoal. Overall, the faunal assemblage indicates the deposition of midden material. There is no evidence of similar material in the 15m or 51m cores, so this midden was presumably limited in extent. There is evidence of further middens on the dune containing the post-Roman industrial structures, only 50m away, where mussel shells are regularly brought to the surface by rabbit activity (Figure 65).

Figure 65. Mussel shells brought to the surface by rabbits burrowing into the post-Roman site.

The midden was covered by marsh [10], with a very large number of molluscs (1780 shells per 500g, 15 taxa) and high diversity (Shannon index 2.34), evenly split between wet ground, catholic and open country taxa, but with no shade species. The wet ground molluscs include *Carychium minimum*, *Galba truncatula*, *Anisus leucostoma*, Succineidae, *Vertigo antivertigo*, *Zonitoides nitidus* and *Ashfordia granulata* as well as one true freshwater mollusc, *Gyraulus crista*, which does not tolerate drying. There are good numbers of catholic (*Cochlicopa* spp., *Trochulus* sp.) and open country (*Vertigo pygmaea*, *Vallonia excentrica*, *Cochlicella acuta*, *Cernuella virgata*, *Helicella itala*) species. This very varied assemblage shows that the local habitat was likely to be swampy with pools of open water and areas of marsh, but with some permanently dry ground. The core position was probably at the margin of wet and dry ground, with grazed land uphill leading to the meadows below the scarp of the towans and marsh with some standing water extending downhill in the river valley. The GPR scan (Figure 44) reflects both the upper and lower limits of this horizon.

A further deposit of windblown sand [11] covers the marsh. Molluscs are scarce implying instability, those present all being open country species indicating that the ground is now dry with no evidence of marshland. The topsoil [12] is more stable, with open country molluscan species. The likely medieval introduction, *Candidula intersecta*, is present in this horizon but at no lower level. The topsoil sample, at 15cm thick, contains fewer molluscs than the modern 4–5cm turf sample, both in numbers and species, indicating that true stability was only established in the very recent past.

51M CORE

Grid ref.: SW 58886.42085; modern ground level: 5.61m OD.
Mollusc numbers: Table 8; mollusc diagram: Figure 66; summary diagram: Figure 67.

Table 8. Mollusc numbers for the 51m core.

Depth in core (cm)	9-21	21-26	26-29	29-34	34-38	38-50	50-160	160-280	280-386	386-388	388-393	393-476	476-588	588-677	677-725	725-729	729-750
Corrected depth (cm)	**0-21**	**21-26**	**26-29**	**29-34**	**34-38**	**38-50**	**50-160**	**160-280**	**280-386**	**386-388**	**388-393**	**393-405**	**476-588**	**588-677**	**677-725**	**725-729**	**729-769**
Sample weight (g)	90	128	30	52	95	251	2460	3054	2645	67	127	2143	2980	2461	801	89	245
Munsell number	7.5YR 3/1	7.5YR 3/2	7.5YR 3/2	7.5YR 4/3	7.5YR 4/3	10YR 6/3	10YR 7/3			10YR 5/2	10YR 6/2	10YR 7/3	10YR 5/2	10YR 7/4	10YR 7/2	gley1 5/N	5YR 2.5/1
Zone	**9**	**8**	**7**			**6**				**5**		**4**				**3**	**2**
Sediment	topsoil	wall debris	marsh			sand				buried soil		sand				gleyed sand	marsh
Carychium minimum	33		40	20													
Carychium sp.									1	1							
Galba truncatula	3	12	48	22	215	10				2				1			
Lymnaea fusca/palustris			1		3												
Cochlicopa lubrica	3																
Cochlicopa sp.									1	4	1		2				
Pupilla muscorum									3	1							
Lauria cylindracea	1																
Vallonia excentrica	42	5	41	37	137		1		1	28	9		3	2			
Punctum pygmaeum											1						
Oxychilus cellarius	1																
Milacidae	1				1					1	1						
Limacidae	1		1														
Cochlicella acuta	3		3	5	16	1		2	7	2	3		3	2			
Ashfordia granulata				1						19			3				
Candidula intersecta	2																
Cernuella virgata	6	2	2	5	5		2	2	3								
Helicella itala			2		2			1						1			
Xerocrassa geyeri													3	1			
Trochulus hispidus				6						1							
Cepaea nemoralis	1			1													
Cornu aspersum	1			1													
TOTAL SHELLS	**98**	**19**	**138**	**98**	**379**	**11**	**3**	**5**	**15**	**59**	**15**	**0**	**14**	**7**	**0**	**0**	**0**
Shells per 500g	544	74	2300	942	1995	22	1	1	3	440	59		2	1			
TOTAL SPECIES	**13**	**3**	**8**	**9**	**7**	**2**	**2**	**3**	**6**	**9**	**5**		**5**	**5**			
Mytilus sp.										+							
Seeds					+												
Wood																+	+
Iron fragments			+														

The position of the 51m core is close to the north side of the 'Cornish hedge' running along the northern side of the B3301 road (Figure 68). Cornish hedges consist of stone walls on either side with infilling of earth. The height of this hedge (140cm) is such that the ground close to its north face is continuously in shade with only a little evening sunlight. Tall vegetation between it and the core site consists largely of tall grasses, rosebay willow herb (*Chamerion angustifolium*), rough chervil (*Chaerophyllum temulum*), mare's tail (*Hippuris vulgaris*) and dock (*Rumex* sp.). There is a slight depression in the ground surface at this point, such that it remains waterlogged with standing water for much of the winter (Charles Thomas, pers. com.), allowing the mare's tail to thrive. This wetness is reflected in the molluscan fauna, where *Carychium minimum* accounts for 49% of all shells present in the turf, and there is a single shell of *Galba truncatula*, a marsh species tolerant of regular moderate desiccation. There are good numbers of *Vallonia excentrica*, normally a short grassland species although occasionally found in wetter situations (Evans 1972, 162), and *Trochulus hispidus* (a catholic species), but only small numbers of other taxa and it is possible that some or all of these dry ground shells are allochthonous, having been blown or washed into this low-lying area from elsewhere.

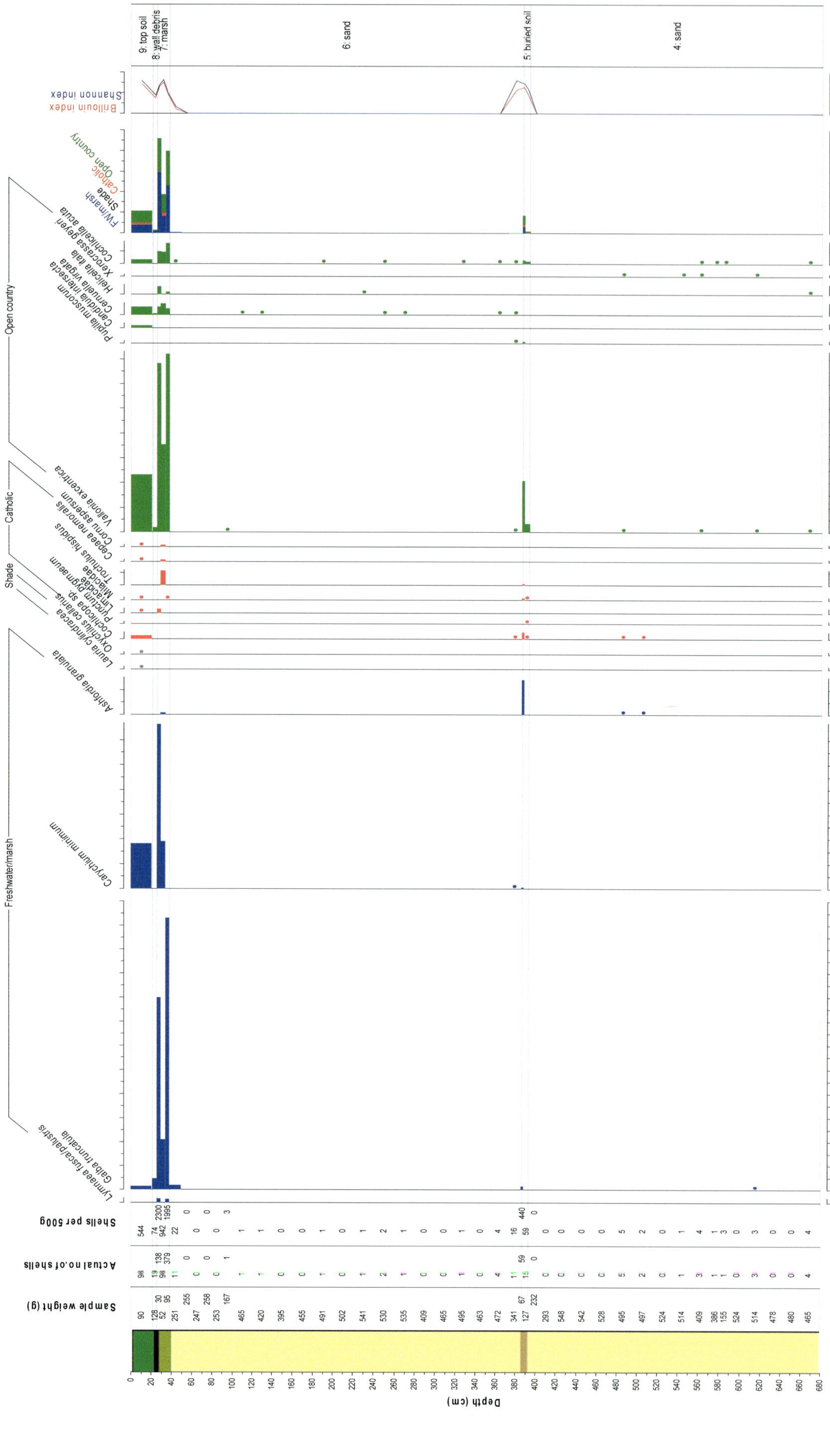

Figure 66. Mollusc diagram for the 51m core. There were no molluscs below 677cm.

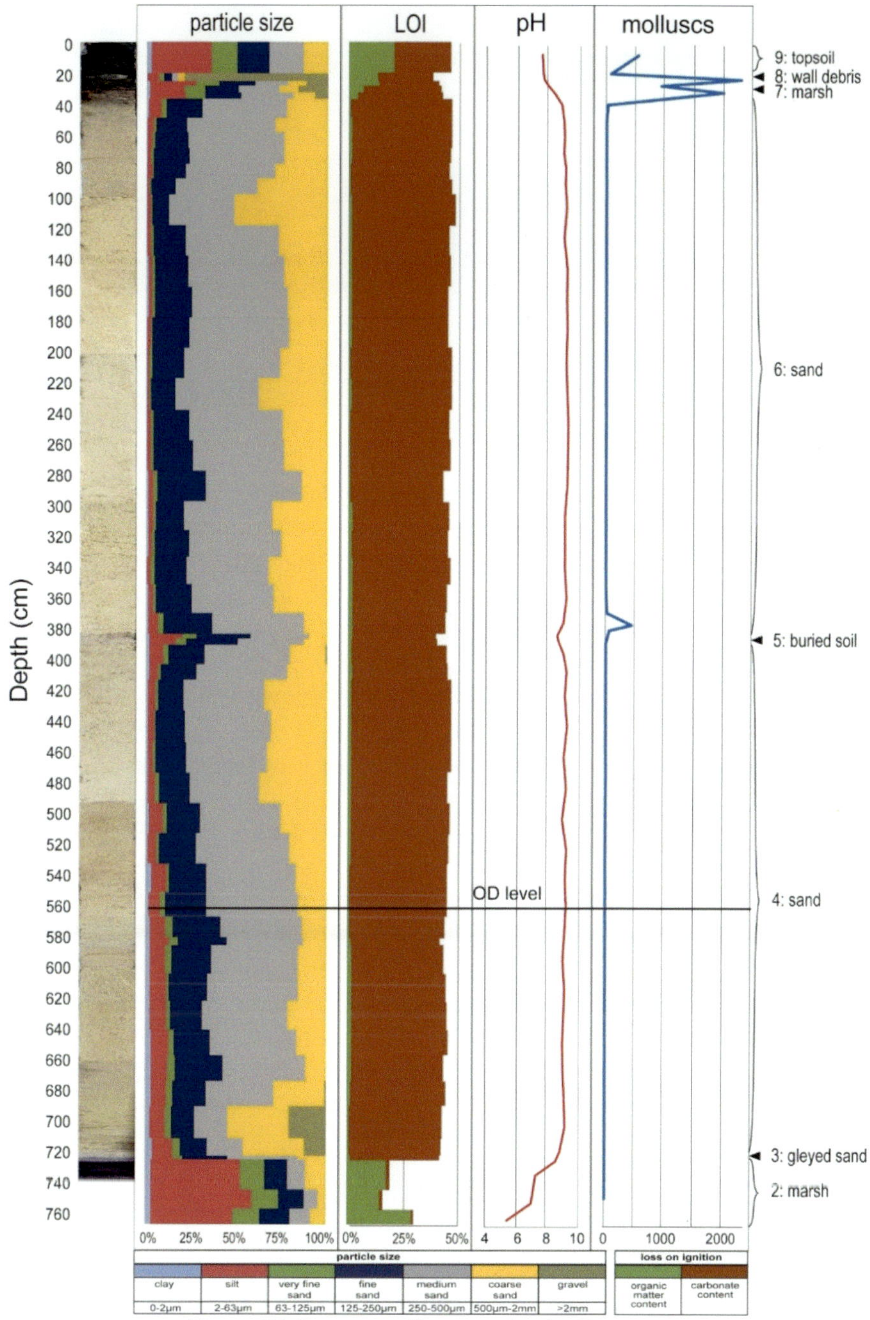

Figure 67. Summary diagrams for the 51m core.

Bedrock was not reached in this core. The lowest sediment obtained was organic silt [2] at 769cm, with an acidic pH (down to 5.3) which almost certainly corresponds to the region of low pH found in the 30m core. The silt is overlain by a thick (336cm) sand layer, the lowest 4cm of which is gleyed [3] while the remainder is yellow sand [4]. This sand contains only very sporadic molluscs of all habitat types. Th3 sparse assemblage suggests that the sands were unstable most of the time during accumulation and these few shells may be allochthonous.

Figure 68. The 51m location (white peg) showing the Cornish hedge and tall vegetation.

A probable 7cm buried soil with higher silt levels is present at 386–393cm [5], corresponding to zone [7] in the 30m core and representing a phase of stabilisation. A moderate number of molluscs is present, with a mixture of wet ground, catholic and open country species, although notably there are no *Lauria cylindracea* in this horizon, only 20m from the location where they were the dominant taxon; this can be explained by the hypothetical walls above being limited to the river bank.

The buried soil is covered by another thick (348cm) layer of blown sand [6] which is almost devoid of molluscs implying continual instability. This sand is capped by a more silty sediment [7] containing very large numbers of wet ground molluscs (up to 2300 per 500g), notably *Galba truncatula* and *Carychium minimum*, as well as grassland species, mainly *Vallonia excentrica*. The assemblage suggests that this land was swampy but subject to drying and is very similar to the modern turf, showing that there has been little change in the local conditions in the recent past. The surrounding slightly higher ground was more continuously dry and probably grazed, which may well account for the open country species.

The 21–26cm horizon in this core [8] is different from any other near-surface horizon in that it contains a large proportion of stones, some of which are fragments of slate cut by the core tube (Figure 69). Molluscs are few and it seems almost certain that the stones are debris from the building of the adjacent Cornish hedge when the road was stabilised, probably in the mid/late nineteenth century.

Figure 69. The upper 44cm of the 51m core showing the slate fragments from the Cornish hedge.

The topsoil [9] contains a moderate number of molluscs of both wet ground (mainly *Carychium minimum*) and open country (mainly *Vallonia excentrica*) taxa consistent with the present regular waterlogging, but adjacent to grazed grassland.

In summary, this core, with marsh only in its lowest levels, indicates that this location was close to the northern limit of the wet river basin. It is almost entirely sand, containing virtually no molluscs implying continual instability over a long period of time, with only one brief episode of greater stability. Sufficient silt to permit a rich molluscan fauna to develop only accumulated in very recent times, and with a significant wet ground component consistent with localised regular ponding.

76m core

Grid ref.: SW 58886.42110; modern ground level: 7.12m OD.
Mollusc numbers: Table 9; mollusc diagram: Figure 71; summary diagram: Figure 72.

The 76m core was obtained part-way up the gentle foot-slope of the west end of the large sand dune on which the post-Roman buildings excavated by Charles Thomas in the 1950s are situated (Figure 70). The core location was on the rise of ground near the terminus of the dune at an elevation of around 1m above the surrounding pasture, and is currently vegetated with short grasses as well as some lady's bedstraw (*Galium verum*), wild thyme (*Thymus polytrichus*), sheep's fescue (*Festuca ovina*) and moss. The turf molluscs are almost entirely open country species as might be expected. Coring was only successful to 760cm as the sand at this depth was so wet that it was not retained in either core tubes or gouges, the latter being used below 540cm. Depths containing organic silt, if they ever extended this far from the river, were not reached. Of importance is that, at least in the area investigated at the westernmost end of the dune, this very prominent mound in the landscape appears to be sand throughout, with no evidence that it has accumulated on a rocky spur.

Figure 70. The sand dune with the post-Roman settlement viewed from the north-west. The location of the 76m core is marked by a red dot.

Table 9. Mollusc numbers for 76m core.

Depth in core (cm)	0-15	15-43	43-56	56-149	149-224	224-262	262-300	310-343	343-453	460-560	560-670	670-760
Corrected depth (cm)	**0-15**	**15-43**	**43-56**	**56-140**	**140-220**	**220-260**	**260-300**	**300-337**	**337-450**	**450-552**	**552-662**	**662-760**
Sample weight (g)	100	74	54	1522	1697	973	931	780	2785	2310	1304	2209
Munsell number	10YR 5/3	10YR 6/3		10YR 6/4				10YR 5/3	10YR 6/4		10YR 7/3	
Zone	**6**	**5**		**4**		**3**			**2**			
Sediment	topsoil	stable sand		sand		stable sand			sand			
Cochlicopa lubrica				1	1	2	1	1				
Cochlicopa lubricella								1				
Cochlicopa sp.					1	1						
Pupilla muscorum			1	5	7	16	20	35				
Vallonia excentrica		1		2	4	8	10	32		1	1	
Punctum pygmaeum						2						
Vitrina pellucida				1		2		2	1			
Limacidae	1											
Cochlicella acuta	16	15	22	24	14	79	53	155	4	8	1	2
Cornuella virgata	4		4	4	3	8	10	11	1			
Helicella itala				2		14	25	6			1	
Xerovrasssa geyeri												3
Cepaea sp.												1
TOTAL SHELLS	**21**	**16**	**27**	**39**	**30**	**132**	**119**	**243**	**6**	**9**	**3**	**6**
Shells per 500g	105	108	250	13	9	68	64	156	1	2	1	1
TOTAL SPECIES	**3**	**2**	**3**	**7**	**6**	**8**	**6**	**8**	**3**	**2**	**3**	**3**

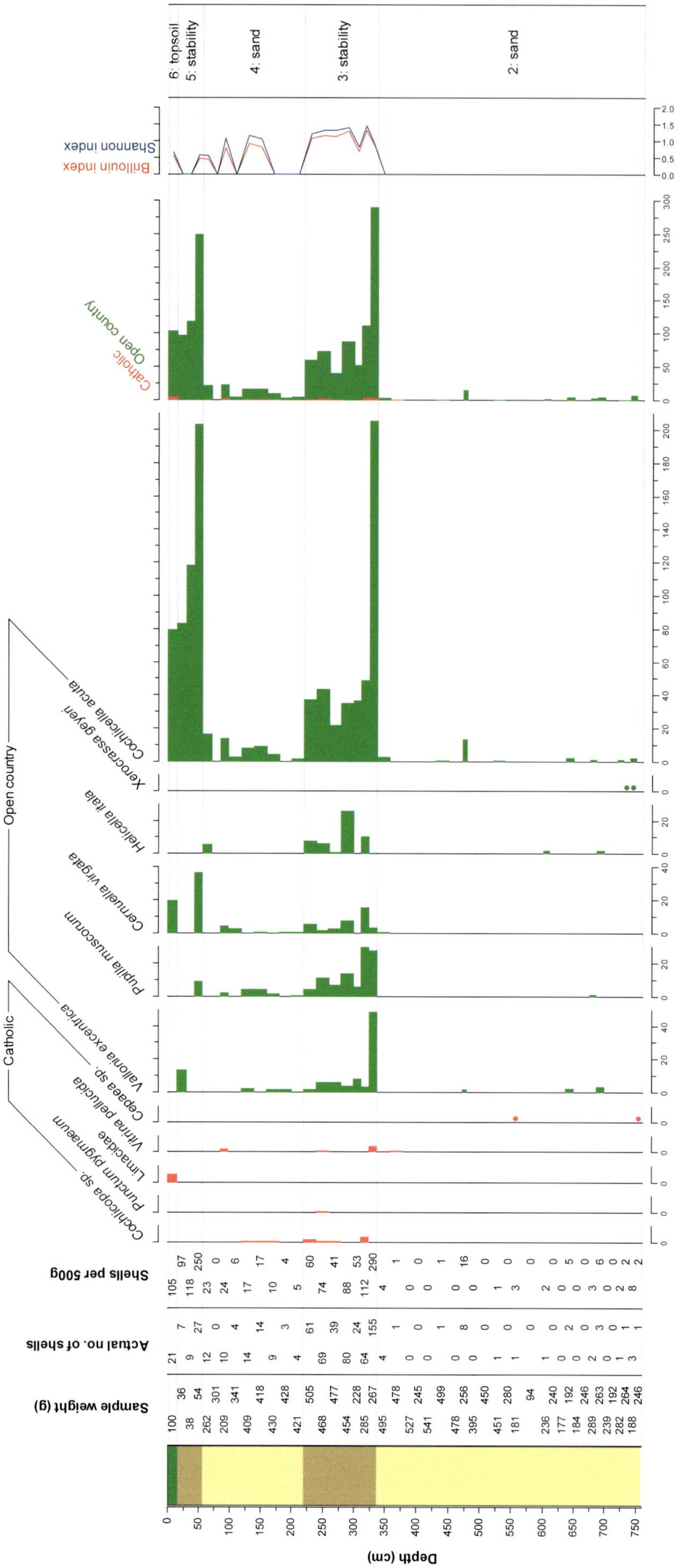

Figure 71. Mollusc diagram for the 76m core. There were no molluscs below 750cm.

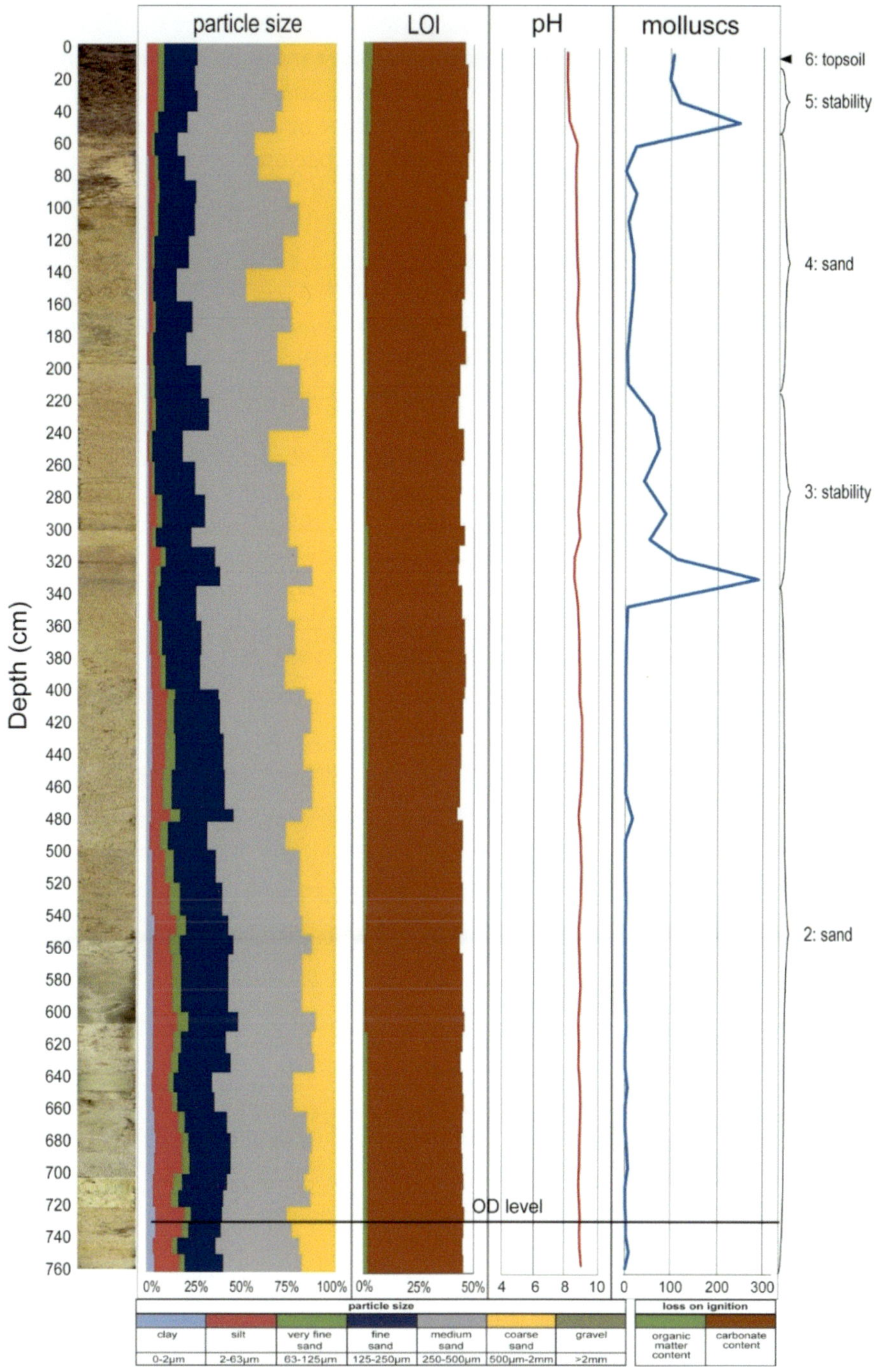

Figure 72. Summary diagrams for the 76m core.

The lower half of this core (337–760cm) is blown sand [2] with only scattered open country shells. There are slightly more molluscs at 474–482cm than elsewhere which may indicate a brief episode of relative stability. There is a very marked increase in the numbers of both shells and taxa at 300-337cm [3], and high numbers of shells persist up to 220cm, although there is no significant change in either the appearance or physical characteristics of the sand indicating that a buried soil did not develop. This molluscan abundance may indicate a period of stability of the sand, but is more likely the result of winnowing. The molluscs are almost entirely xerophilic species with *Cochlicella acuta* strongly predominant, with *Pupilla muscorum, Vallonia excentrica*, *Cernuella virgata* and *Helicella itala* also present, as well as a few with more catholic tolerances, but no taxa requiring shade. The strong presence of *Vallonia* suggests that the ground was grazed; this assemblage is very similar to that of the modern short grazed grass. If the Romano-British date for the introduction of *Cernuella* is accepted (Kerney 1999, 181) then its presence down to a depth of 330cm indicates that much of the sand at this end of the dune has accumulated during historic times, and the extent of the dune may have been much smaller during the early medieval period.

Instability returns with further blown sand [4] containing few shells; this is overlain by another period of stability [5], although with far fewer molluscs than in the lower mollusc-rich level, almost all of which are open country taxa. The topsoil [6] also contains open country species as well as a single Limacidae plate. These upper layers are consistent with the bare grassland now present, but the absence of *Vallonia* in the topmost 15cm, although present in the turf sample, implies that only recently has this dune returned to short grazed grass.

The meadow – 100m to 245m

The cores from 100m to 245m were obtained across the gently sloping meadow between the scarp of Godrevy Towans and the B3301 road (Figure 73). The ground is much disturbed by rabbit burrows, and the meadow is regularly grazed by cattle and sheep. A wide variety of meadow plants is present:

- Dock (*Rumex* sp.)
- Creeping cinquefoil type (*Potentilla reptans*)
- Yellow vetchling (*Lathyrus aphaca*)
- Rest harrow (*Ononis spinosa*)
- Bird's foot trefoil (*Lotus corniculatus*)
- Clover (*Trifolium* sp.)
- Lady's bedstraw (*Galium verum*)
- Wild thyme (*Thymus polytrichus*)
- Ribwort plantain (*Plantago lanceolata*)
- Creeping thistle (*Cirsium arvense*)
- Smooth sow-thistle (*Sonchus oleraceus*)
- Few-leafed hawkweed (*Hieracium umbellatum* agg.)
- Sea couch grass (*Elytrigia juncea*)
- Sheep's fescue (*Festuca ovina*)
- Unidentified mosses

Figure 73. View across the meadow with the slope of Godrevy Towans in the background.

The modern turf molluscs across the meadow show a fairly consistent pattern with a good number of shells, nearly all of which are those associated with this type of pasture. *Vallonia excentrica* is dominant at almost all locations, but with *Vertigo pygmaea*, *Cochlicella acuta* and *Helicella itala* strongly represented. These assemblages are entirely consistent with the grazed grassland of the meadow which abounds with rabbit burrows and is regularly browsed by sheep and cattle. The abundance of *V. excentrica* (50% of all shells) supports the hypothesis that high proportions of *Vallonia* are consistent with established grazed grassland (Cameron and Morgan-Huws 1975; Evans and Evans 1995).

100M CORE

Grid ref.: SW 58886.42140; modern ground level: 6.35m OD.
Mollusc numbers: Table 10; mollusc diagram: Figure 74; summary diagram: Figure 75.

Table 10. Mollusc numbers for the 100m core.

Depth in core (cm)	0-12	12-100	113-124	124-220	220-334	334-345	345-400	400-489	489-500
Corrected depth (cm)	**0-12**	**12-100**	**108-117**	**117-220**	**220-334**	**334-345**	**345-400**	**400-489**	**489-500**
Sample weight (g)	111	1725	207	2073	2633	290	1518	2388	430
Munsell number	7.5YR 3/2	10YR 6/4	7.5YR 2.5/2	10YR 6/4					10YR 3/2
Zone	**7**	**6**	**5**	**4**		**3**	**2**		**1**
Sediment	topsoil	sand	buried soil	sand		buried soil	sand		killas
Cochlicopa spp.								1	
Galba truncatula		1		1					
Pupilla muscorum		2		2		4	1	1	
Lauria cylindracea								1	
Vallonia excentrica	9		3	1		1	3	6	
Discus rotundatus						2		1	
Aegopinella nitidula							1		
Clausilia bidentata							3	2	
Cochlodina laminata								1	
Cochlicella acuta	44	21	4	7	3	15	3	9	
Candidula intersecta	12	7	1		1				
Cernuella virgata	3	1	2	2					
Helicella itala		1		2					
Xerocrassa geyeri							1	1	
Cepaea sp.		2			1			1	
Cornu aspersum			1						
TOTAL SHELLS	**68**	**35**	**11**	**15**	**5**	**22**	**12**	**24**	**0**
Shells per 500g	306	10	27	4	1	38	4	5	
TOTAL SPECIES	**4**	**7**	**5**	**6**	**3**	**4**	**6**	**9**	
Mytilus sp.		+							
Ostrea sp.		+							
Charcoal				+					

The 100m core reached bedrock [1] at a depth of 489cm, well seen on the GPR (Figure 44), and overlain with sand [2] containing an assemblage of mollusc species of varying habitat requirements. Of interest is the presence of a few *Discus rotundatus*, *Clausilia bidentata* and *Cochlodina laminata*, all species associated with leaf litter and shade, perhaps suggesting some woodland or scrub in the near vicinity during the time that part of the sequence was deposited. While the shade-loving molluscs may be intrusive due to bird or animal movement their association and the lack of any wet ground species makes an autochthonous explanation more likely. *Clausilia* was found in the modern turf samples at 30m (long grass), at 345m (bare ground) and at 400m (marram grass) but *Discus* and *Cochlodina* were not present in any turves. Open country species are also present here indicating that there was probably open scrub nearby.

A stabilisation horizon with buried soil [3] is indicated by the slight increase in silt proportions and of molluscs, the latter being mostly open country taxa, but again with two specimens of *Discus rotundatus* suggesting that some wood/scrub persisted in the area. A possible interface on the GPR corresponds to this horizon (Figure 44). Deep sand [4] covers the soil which contains very few shells; a single *Galba truncatula*, a wet ground mollusc, was probably deposited here by wind or birds. This sand is overlain by a buried soil [5], with increase in clay, silt and organic matter. Shells remain few and are all of open country species apart from a single catholic shell, *Cornu aspersum*.

Further blown sand [6] represents another period of instability with few shells except in its uppermost portion where there is an increase in xerophilic species, especially *Cochlicella acuta* and *Candidula intersecta*. Another *Galba truncatula* is almost certainly allochthonous. The topsoil [7] contains solely open country taxa, with *Vallonia excentrica* reappearing consistent with the current grazed grassland on the site.

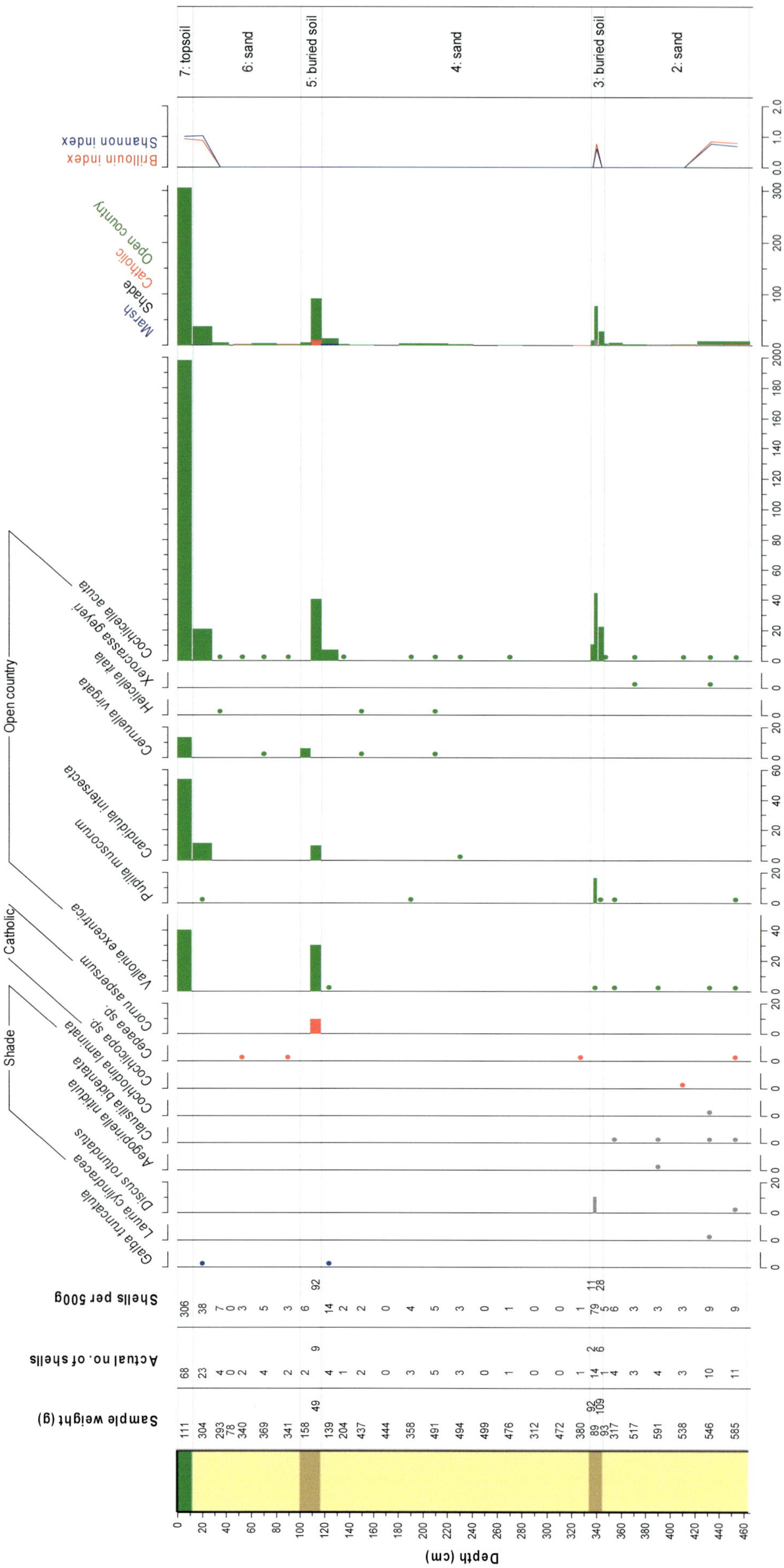

Figure 74. Mollusc diagram for the 100m core. There were no molluscs below 462cm.

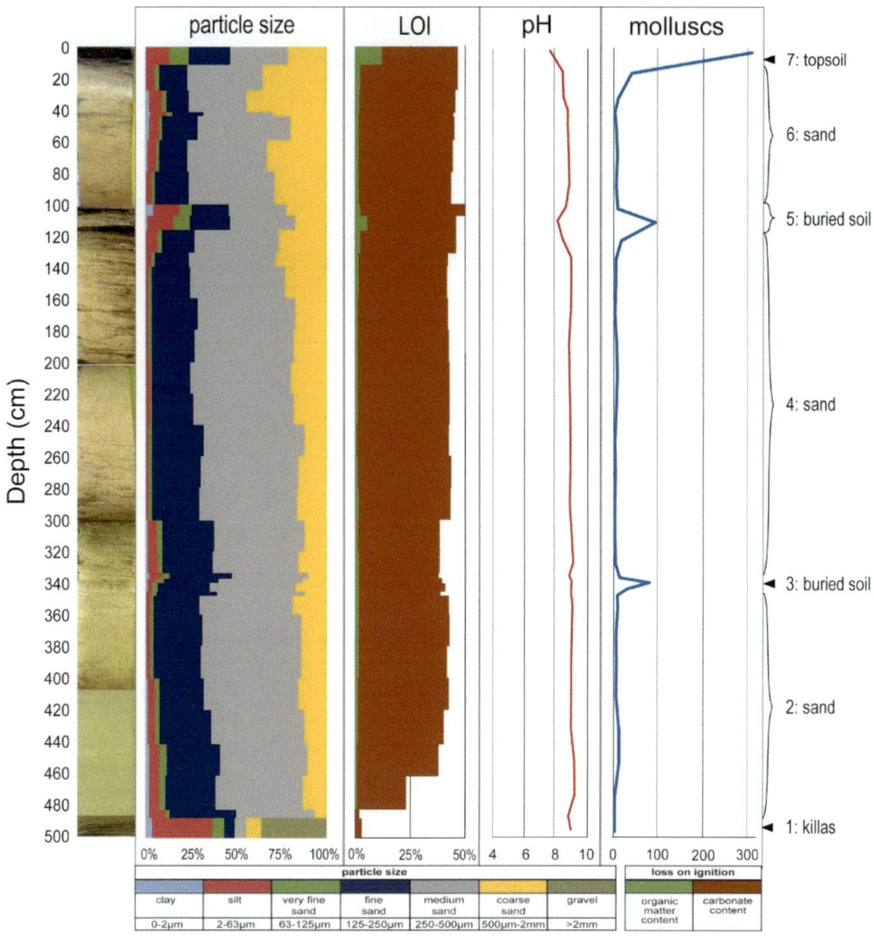

Figure 75. Summary diagrams for the 100m core.

125M CORE

Grid ref.: SW 58886.42159; modern ground level: 6.34m OD.
Mollusc numbers: Table 11; mollusc diagram: Figure 76; summary diagram: Figure 77.

Killas [1] was found at a depth of 175cm, consistent with the GPR (Figure 44) which indicates that this location is at the margin of the original river bank. This contained some mussel (*Mytilus* sp.), wood and charcoal fragments, almost certainly displaced from overlying sediments or in a feature fill containing redeposited bedrock, and indicative of human presence. 62cm of blown sand [2] covers the killas with no trace of any buried soil. A few shells are present in this sand, entirely *Xerocrassa geyeri* apart from a single *Aegopinella nitidula*, indicating open ground. This sand is overlain by horizon [3] which has a slight increase in silt at 100–113cm, implying a stabilisation zone with a buried soil. It contains a high number of shells, nearly all open country taxa, and is the highest level containing *X. geyeri*, strongly supportive of this being a Bronze Age horizon (see Chapter XIV for discussion of the extinction of *X. geyeri*). The catholic *Cochlicopa lubrica* is well represented and there are single specimens of *Clausilia bidentata* and *Galba truncatula*.

The sand overlying this buried soil [4] contains a few more molluscs than at 100m, all open country species, with high numbers of *Cochlicella acuta* in the uppermost portion. The appearance of *Cernuella virgata*, a Romano-British introduction to Britain (Kerney 1999, 181), shows that this sand was deposited or reworked in historic times. The topsoil [5] is rich in shells, again all open country taxa except for three *Carychium minimum* and one *Galba truncatula*, wet ground species which are probably allochthonous. *Vallonia excentrica* is abundant while there is only a single specimen of *Pupilla muscorum*, entirely consistent with grazed grassland.

Table 11. Mollusc numbers for the 125m core.

Depth in core (cm)	0-10	10-20	20-30	30-40	40-50	50-60	60-70	70-80	80-90	90-100	120-130	130-140	140-150	150-160	160-170	170-181	181-190	190-200
Corrected depth (cm)	**0-10**	**10-20**	**20-30**	**30-40**	**40-50**	**50-60**	**60-70**	**70-80**	**80-90**	**90-100**	**100-113**	**113-125**	**125-138**	**138-150**	**150-163**	**163-175**	**175-188**	**188-200**
Sample weight (g)	55	119	195	209	214	231	251	140	234	213	231	255	260	258	256	303	268	257
Munsell number	7.5YR 3/2	7.5YR 3/2	7.5YR 5/2	10YR 7/4	10YR 7/4	10YR 7/4	10YR 7/4	10YR 7/4	10YR 7/4	10YR 7/4	10YR 7/4	10YR 7/4	10YR 7/4	10YR 7/4	10YR 7/4	10YR 7/4	10YR 5/3	10YR 4/4
Zone	5		4								3	2					1	
Sediment	topsoil		sand								buried soil	sand					killas	
Carychium minimum		3									1							
Galba truncatula	1																	
Cochlicopa lubrica											12							
Vertigo pygmaea	1	1																
Pupilla muscorum		1			1			1	2	1	25							
Vallonia excentrica	20	54	1	1		1	1	1			17							
Clausilia bidentata											1							
Aegopinella nitidula															1			
Cochlicella acuta	15	121	46	5	2	4	4	4	4		12							
Candidula intersecta	2	3																
Cernuella virgata	9	25	8		1	1	2											
Helicella itala	5	12	2		1	5	3	3	6									
Xerocrassa geyeri											15	2	3	1	4	1	2	
Cepaea sp.						1												
TOTAL SHELLS	**53**	**220**	**57**	**6**	**5**	**12**	**10**	**9**	**12**	**1**	**83**	**2**	**3**	**1**	**5**	**1**	**2**	**0**
Shells per 500g	482	924	146	14	12	26	20	32	26	2	180	4	6	2	10	2	4	
TOTAL SPECIES	**7**	**8**	**4**	**2**	**4**	**5**	**4**	**4**	**3**	**1**	**7**	**1**	**1**	**1**	**2**	**1**	**1**	
Mytilus sp.			y													+	+	+
Wood																		+
Charcoal																	+	+

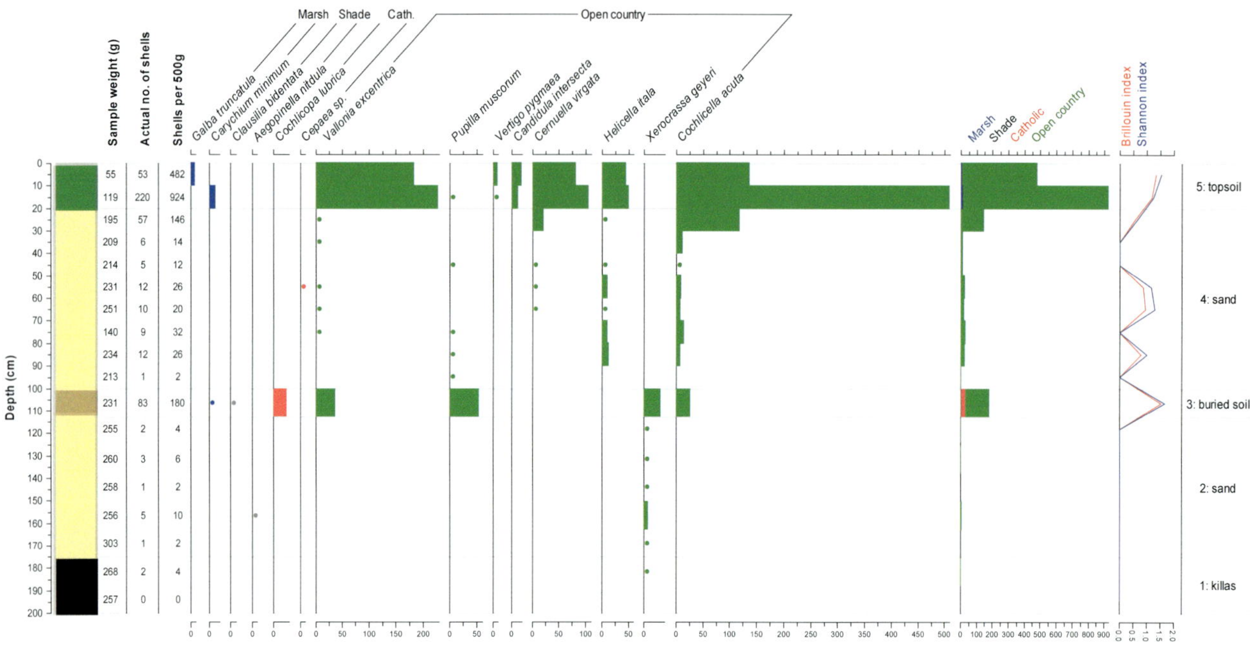

Figure 76. Mollusc diagram for the 125m core.

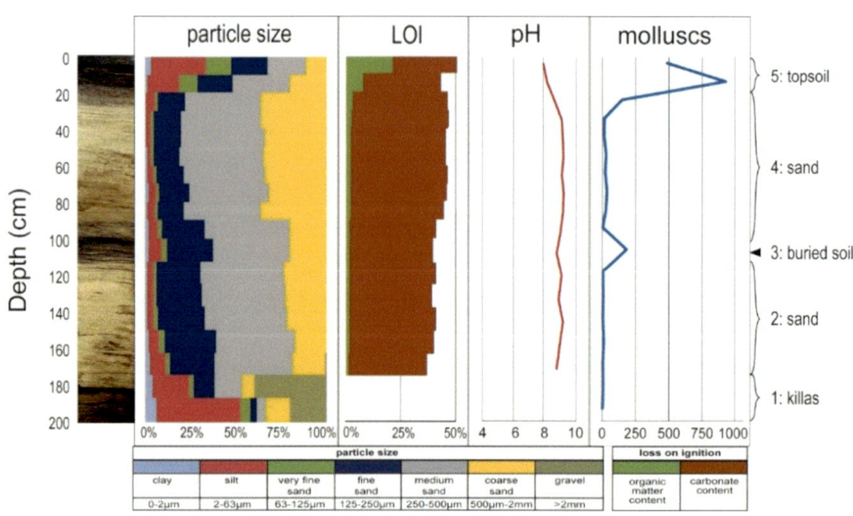

Figure 77. Summary diagrams for the 125m core.

A broken Mesolithic flint blade was found in the basal sediment of this core (Figure 78). Blades and microliths have previously been described from Gwithian (Thomas 1958; Roberts 1987; Thomas 2007), although none are recorded from the area of the later Bronze Age settlement or in the meadow in the region of the coring transect (Sturgess 2004, Fig. 5) .

Figure 78. The flint blade found at the base of the 125m core.

150m core

Grid ref.: SW 15889.42184; modern ground level: 8.03m OD.
Mollusc numbers: Table 12; summary diagram: Figure 79.

Table 12. Mollusc numbers for the 150m core.

Depth in core (cm)	0-17	17-24	24-69	69-82	82-100	105-175
Corrected depth (cm)	**0-17**	**17-24**	**24-69**	**69-82**	**82-100**	**100-175**
Sample weight (g)	144	128	1010	257	349	3107
Munsell number	7.5YR 3/2	10YR 7/4	10YR 7/3	10YR 5/3	10YR 6/4	
Zone	**4**	**3**		**2**		**1**
Sediment	topsoil	sand		buried soil		killas
Cochlicopa lubrica				2		
Vertigo pygmaea	7					
Pupilla muscorum			1	8		
Vallonia excentrica	24		5	31		
Punctum pygmaeum	1					
Cochlicella acuta	65	7	13	23		
Candidula intersecta	3					
Cernuella virgata	8	8	1			
Helicella itala			6			
Xerocrassa geyeri			4	20	8	
Trochulus hispidus	3					
TOTAL SHELLS	**111**	**15**	**30**	**84**	**8**	**0**
Shells per 500g	385	59	15	163	11	
TOTAL SPECIES	**7**	**2**	**6**	**5**		
Mytilus sp.				+	+	
Charcoal				+	+	

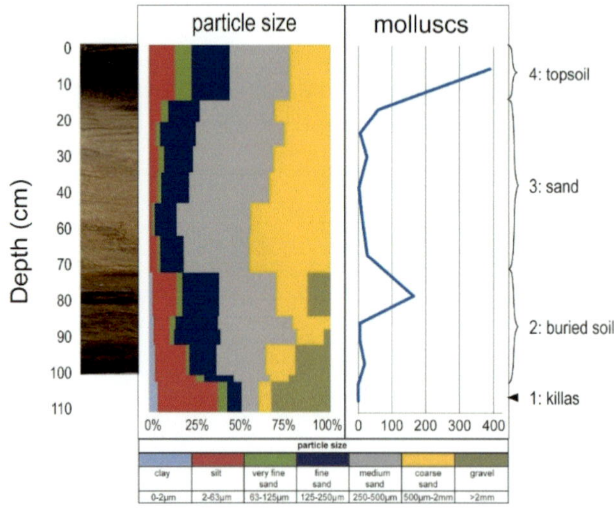

Figure 79. The particle size diagram and mollusc graph for the 150m core.

This core will only be mentioned briefly, as it is the location of the trench which was excavated in 2012 which is fully discussed in the Chapter X.

The percussion core, obtained in 2011, reached a depth of 175cm, with bedrock at approximately 110cm; the exact level was difficult to determine as there was gradual increase in the proportion of gravel below 91cm, unlike other cores where there was a clear-cut change in appearance and sedimentology. What was clear is an increase in silt and in the number of molluscs at 69–82cm, consistent with a buried soil. A further slight increase in molluscs at 91–100cm may indicate a further buried soil. Many fragments of mussel shell and some charcoal were found in the 69–82cm sample.

When the decision was made to excavate a small trench this location was selected partly on the basis of the above findings but also because this position is just outside the area of the post-Roman field system excavated in the 1960s. The core findings will not be discussed further, as considerably more detail was obtained from the trench excavation.

Chapter VII – Hand auger cores

Thirteen hand auger cores were obtained along the transect from the meadow at 175m and the summit of Godrevy Towans at 582m. Initially cores were taken with a 2.5cm gouge and the particle size and loss on ignition measurements were made on those; the cores were later repeated using a 5cm gouge to obtain additional material for mollusc analysis and the mollusc graphs and diagrams use the numbers from those larger cores.

pH graphs have not been included in the summary diagrams for the hand auger cores; the pH for every sample ranged between 7.1 and 8.8, with very little variation within each core. In no sample did pH fall into the acidic range, unlike many of the deeper samples in the percussion cores.

175M CORE

Grid ref.: SW 58886.42209; modern ground level: 9.51m OD.
Mollusc numbers: Table 13; mollusc diagram: Figure 80; summary diagram: Figure 81.

Table 13. Mollusc numbers for the 175m core.

Depth (cm)	**0-9**	**9-15**	**15-20**	**20-25**	**25-31**	**31-37**
Sample weight (g)	245	278	323	332	424	434
Munsell number	10YR 3/2	10YR 4/3	10YR 5/3	10YR 5/4		
Zone	**4**	**3**			**2**	**1**
Sediment	topsoil	buried soil			buried soil	killas
Cochlicopa lubrica	4	12	10	10	2	
Cochlicopa sp.	1	2	5	9		
Vertigo pygmaea	15	1	1			
Pupilla muscorum	4	27	38	37	6	
Vallonia excentrica	145	95	69	44	7	
Punctum pygmaeum	5	1				
Aegopinella nitidula		1				
Limacidae		2				
Cochlicella acuta	78	140	114	76	9	
Candidula intersecta	6					
Cernuella virgata	12	6	22	35		
Helicella itala		27				
Xerocrassa geyeri					29	
Trochulus hispidus	9		2			
Cepaea sp.	1		1		2	
TOTAL SHELLS	**280**	**314**	**262**	**211**	**55**	**0**
Shells per 500g	571	565	406	318	65	
TOTAL SPECIES	**10**	**10**	**8**	**5**	**6**	
Mytilus	+	+	+	+		
Nucella lapillus					+	
Pecten sp.				+	+	
Patella sp.					+	
Charcoal				+	+	

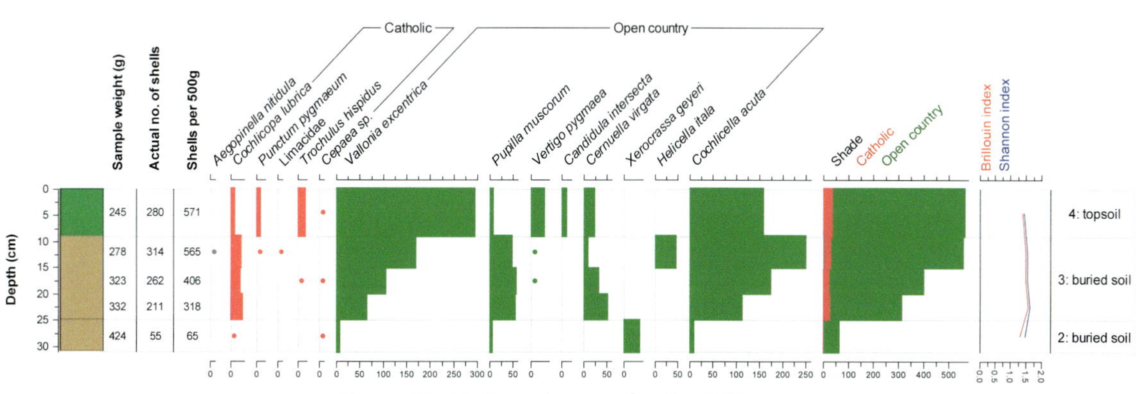

Figure 80. Mollusc diagram for the 175m core.

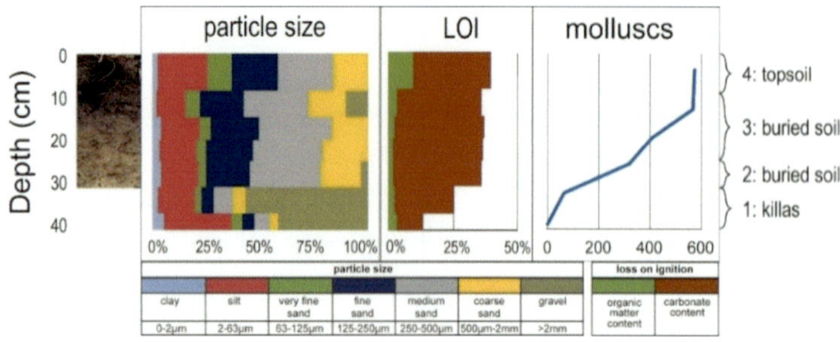

Figure 81. Summary diagrams for the 175m core.

This core is located within the area of the field system in the meadow to the west of the Bronze Age and post-Roman settlements at Gwithian, excavated in the early 1960s (Fowler and Thomas 1962). The core reached the underlying killas [1] at a depth of 31cm with the sediments above this [2, 3] being very consistent in particle size. There were numerous fragments of mussel shells (*Mytilus* sp.) in all levels below 9cm, especially at 25–31cm, together with scallop fragments (*Pecten* sp.) below 20cm and with dog whelk (*Nucella lapillus*) and limpet (*Patella* sp.) shells below 25cm. These findings, together with some charcoal in the lower levels, are consistent with manuring of the fields, probably with seaweed from the beach and some midden material, as proposed by Fowler and Thomas (1962). Mollusc numbers are poor in the lowest sediment [2] but increase considerably above 25cm [3]. *Vallonia excentrica* in particular becomes more and more evident as the sediments accumulate, while there is a corresponding decrease in *Pupilla muscorum*, consistent with change from arable to grazing over time. However considerable caution is needed with this interpretation as ploughing will have resulted in marked bioturbation aided and abetted by rabbit activity. The presence of high numbers of *Xerocrassa geyeri* in level [2] with none higher up indicates that this lowest deposit is relatively undisturbed and reflects Bronze Age activity. *Cernuella virgata*, a post-Roman introduction to Britain, then replaces *Xerocrassa* [3], entirely consistent with these upper levels of the field system being post-Roman in age. *Candidula intersecta*, considered to be a medieval introduction, is only found in the topsoil sample.

Vallonia excentrica is markedly dominant in the topsoil [4], as in the modern turf sample, while *Pupilla muscorum* is hardly represented, and was not found in the turf sample, all consistent with the present grazed grassland.

200m core

Grid ref.: SW 58887.42234); modern ground level: 11.71m OD.
Mollusc numbers: Table 14; mollusc diagram: Figure 82; summary diagram: Figure 83.

Table 14. Mollusc numbers for the 200m core.

Depth (cm)	0-6	6-15	15-22	22-48
Sample weight (g)	241	383	394	1005
Munsell number	10YR 3/2		10YR 6/3	
Zone	**3**		**2**	**1**
Sediment	topsoil		sand	killas
Cochlicopa sp.		1		
Vertigo pygmaea	2	11		
Vallonia excentrica	60	104	1	
Punctum pygmaeum	2	3		
Oxychilus cellarius		1		
Limacidae		1		
Cochlicella acuta	54	95	1	1
Candidula intersecta	7	1		
Cernuella virgata	6	12	1	
Helicella itala	2	7	1	
Xerocrassa geyeri				1
Trochulus hispidus	5	14		
Cepaea sp.	1			
TOTAL SHELLS	**139**	**250**	**4**	**2**
Shells per 500g	288	326	5	1
TOTAL SPECIES	**9**	**11**	**4**	**2**

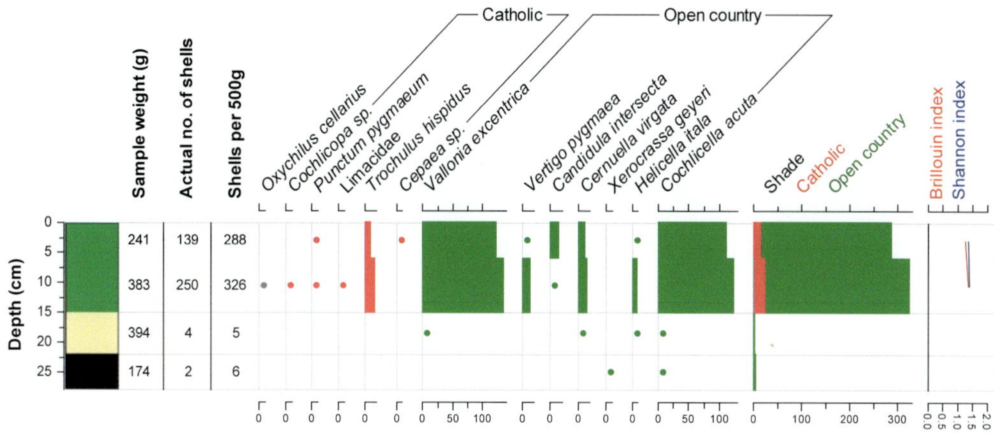

Figure 82. Mollusc diagram for the 200m core.

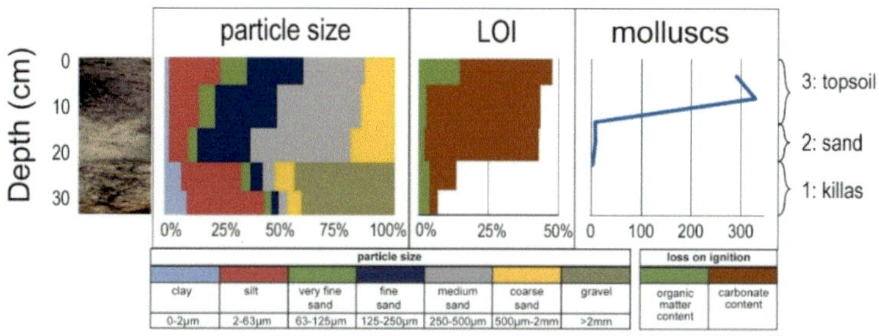

Figure 83. Summary diagrams for the 200m core.

The 200m core is also within the post-Roman field system, and is on the northern edge of a modern grassy trackway running east-west across the meadow. The sand is very shallow, with the killas [1] being reached at 22cm.

The sand [2] contains very few shells but the topsoil [3] is rich in molluscs, mainly xerophilic species, and entirely consistent with the medium height grazed grass now present. Of interest is that no *Pupilla muscorum* were found in these samples, despite the overall large number of shells. *Oxychilus cellarius*, a shade species, can live at the base of vegetation, or could have been deposited at this site by aeolian action or by animals.

Although within the confines of the field system described by Fowler and Thomas (1962) the findings differ from those at 175m, also within the field. Molluscs are only present in any number much nearer the surface of the topsoil, and there were no fragments of marine shells, which had been found at 175m. Whether this area was more unstable than that further from the scarp of the towans or was less intensively used for agricultural purposes so that soils were less able to form, cannot be determined.

225M CORE

Grid ref.: SW 58886.42259; modern ground level: 14.46m OD.
Mollusc numbers: Table 15; mollusc diagram: Figure 84; summary diagram: Figure 85.

Table 15. Mollusc numbers for the 225m core.

Depth (cm)	0-4	4-15	15-23	23-31	31-39	39-48	48-53	53-58	58-62	62-68
Sample weight (g.)	60	375	420	204	169	216	128	118	279	39
Munsell number	10YR 3/2		10YR 4/6	10YR 4/6	10YR 5/4				10YR 4/3	
Zone	4	3			2					1
Sediment	topsoil	sand								killas
Cochlicopa lubrica	3		4							
Cochlicopa sp.	1		3	1	1					
Vertigo pygmaea	6	18								
Pupilla muscorum			21	4	1					
Vallonia excentrica	34	132	11	5	1					
Punctum pygmaeum		4								
Aegopinella nitidula	1									
Cochlicella acuta	3	132	46	34						
Candidula intersecta		20	23							
Cernuella virgata	2	28		9	5					
Helicella itala		1	2							
Xerocrassa geyeri						8	1			
Trochulus hispidus		7								
Cepaea sp.		1								
TOTAL SHELLS	**50**	**343**	**110**	**53**	**8**	**8**	**1**	**0**	**0**	**0**
Shells per 500g	417	457	131	130	24	19	4			
TOTAL SPECIES	**6**	**9**	**6**	**5**	**4**	**1**	**1**			
Mytilus sp.		+	++	++	+	+++	++	+++	+	
Charcoal							+			

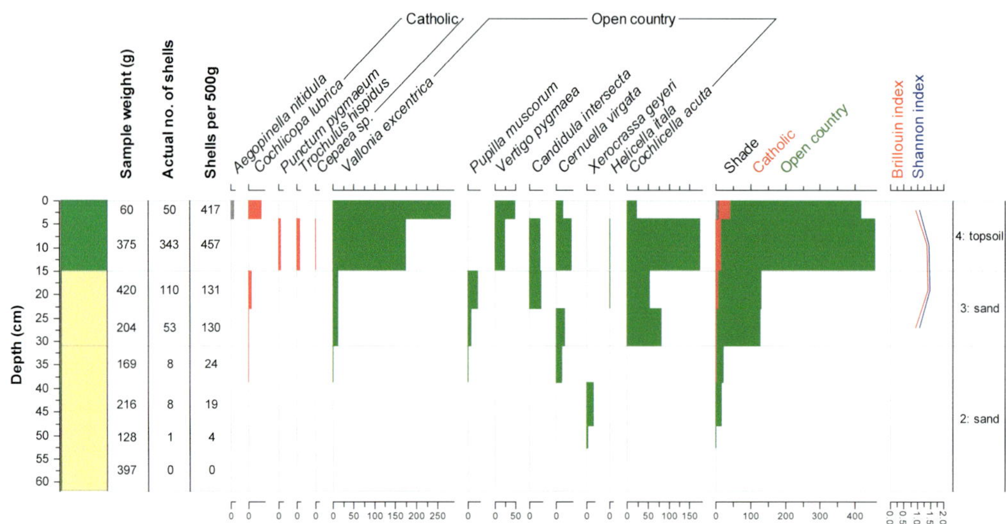

Figure 84. Mollusc diagram for the 225m core.

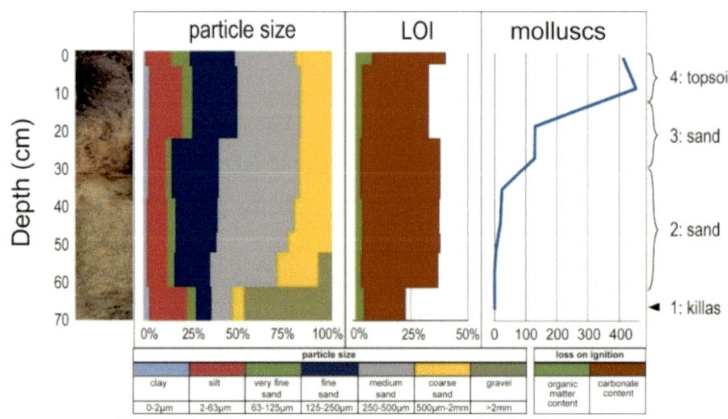

Figure 85. Summary diagrams for the 225m core.

This core was obtained from an area of sloping meadow close to the base of the scarp rising to the summit of the towans. Killas [1] was reached at 62cm, considerably deeper than at 200m and perhaps where some of the colluvial sediment from the scarp has accumulated. The lower layer of sand [2] is almost devoid of shells but does include *Xerocrassa geyeri*, implying that this layer has been little disturbed. Mollusc numbers increase considerably in the sand from 31cm upwards [3]. *Pupilla muscorum*, a species almost absent from other meadow cores, is found, particularly at 15–23cm, showing that this area was probably less grazed than further down the meadow, perhaps due to bare ground at the base of the towans scarp (Evans 1972, 146). In the topsoil [4] it is replaced by *Vallonia excentrica* – also present in high numbers in the modern turf block – showing more stability of the ground with grazing. Other topsoil shells are nearly all open country species but with some more catholic taxa including a single *Aegopinella nitidula* which can live in the damp shade at the base of meadow grasses.

245m core

Grid ref.: SW 58887.42279; modern ground level: 17.02m OD.
Mollusc numbers: Table 16; mollusc diagram: Figure 86; summary diagram: Figure 87.

Table 16. Mollusc numbers for the 245m core.

Depth (cm)	0-10	10-18	18-30
Sample weight (g)	236	446	509
Munsell number	10YR 3/2	10YR 4/4	10YR 6/4
Zone	**3**	**2**	
Sediment	topsoil	sand	
Cochlicopa sp.	5		
Vertigo pygmaea	3		
Pupilla muscorum	2		
Vallonia excentrica	41		
Cochlicella acuta	41	3	
Candidula intersecta	7	1	
Cernuella virgata	10	3	
Xerocrassa geyeri			
Trochulus hispidus	3		
Cornu aspersum	1		
TOTAL SHELLS	**113**	**7**	**0**
Shells per 500g	239	8	
TOTAL SPECIES	**9**	**3**	

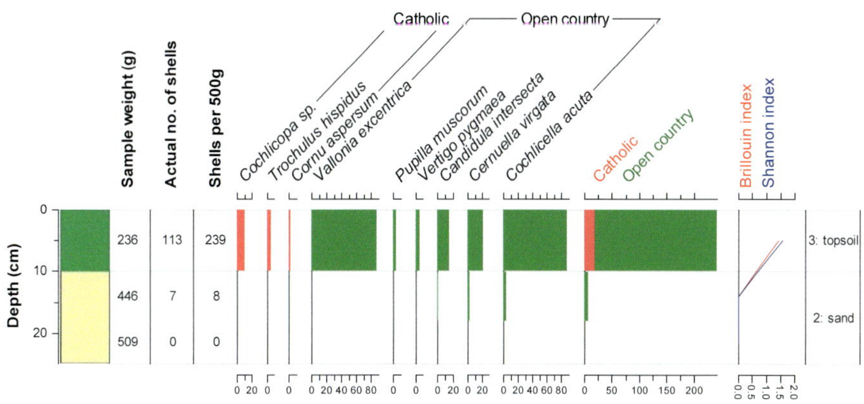

Figure 86. Mollusc diagram for the 245m core.

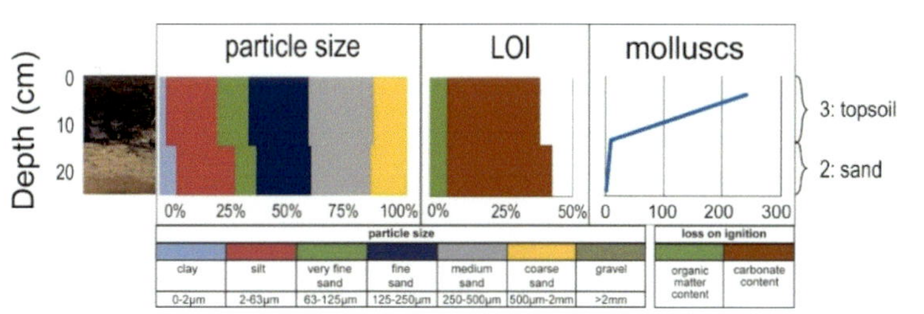

Figure 87. Summary diagrams for the 245m core.

The 245m core was taken as close as possible to the base of the steep scarp rising to the summit of Godrevy Towans (Figure 88). It was within 0.5m of blackthorn (*Prunus avium*) scrub which covers the lower slopes of the scarp; the aspect is directly south facing although the scrub provides some shade. The sand overlying the killas is very thin, the latter being reached at only 25cm. The sand [2] contains a moderate quantity of silt, as might be expected from run-off from the scarp, but there are virtually no molluscs, implying it was unstable, being regularly removed and replaced by wind/water action. The topsoil [3], in contrast, contains a good number of shells indicating relative stability of the sediments in more modern times.

Figure 88. The 245m core position (white peg) at the edge of the scrub at the base of the Towans scarp.

289M CORE

Grid ref.: SW 58887.42324; modern ground level: 33.71m OD.
Mollusc numbers: Table 17; mollusc diagram: Figure 89; summary diagram: Figure 90.

Table 17. Mollusc numbers for the 289m core.

Depth (cm)	0-20	20-38	38-44	44-50	50-60	60-70	70-80	80-90	90-100	100-120	120-140	140-160	160-180	180-200	200-220	220-240	240-260	260-280	280-300	300-320	320-335
Sample weight (g)	48	79	50	62	107	101	93	92	39	89	81	78	189	173	181	218	213	200	226	204	63
Munsell number	10YR 5/3		7.5YR 5/3	7.5YR 6/4	10YR 7/4																
Zone	**5**	**4**	**3**	**2**																	
Sediment	topsoil	sand	buried soil	sand																	
Cochlicopa sp.												1									
Vertigo pygmaea			2																		
Pupilla muscorum	2	1	1		1	2	6	4			1		5	8	4			7	1		
Lauria cylindracea			5								1										
Vallonia excentrica	1	1																			
Aegopinella nitidula																					
Cochlicella acuta			3		1	3	3	4		2	3		5	1	3		1	2		2	1
Candidula intersecta				1	1							1									
Cernuella virgata											1	1	1								
Helicella itala	3	2	3		2	3	2	4	1		1	2	7	3	5	8	4	5	6	3	2
Cornu aspersum				1		1															
TOTAL SHELLS	**6**	**4**	**14**	**2**	**5**	**9**	**11**	**12**	**1**	**2**	**7**	**5**	**18**	**12**	**12**	**8**	**5**	**14**	**7**	**5**	**3**
Shell per 500g	63	25	140	16	23	45	59	65	13	11	43	32	48	35	33	18	12	35	15	12	24
TOTAL SPECIES	**3**	**3**	**5**	**2**	**4**	**4**	**3**	**3**	**1**	**1**	**5**	**4**	**4**	**3**	**3**	**1**	**2**	**3**	**2**	**2**	**2**

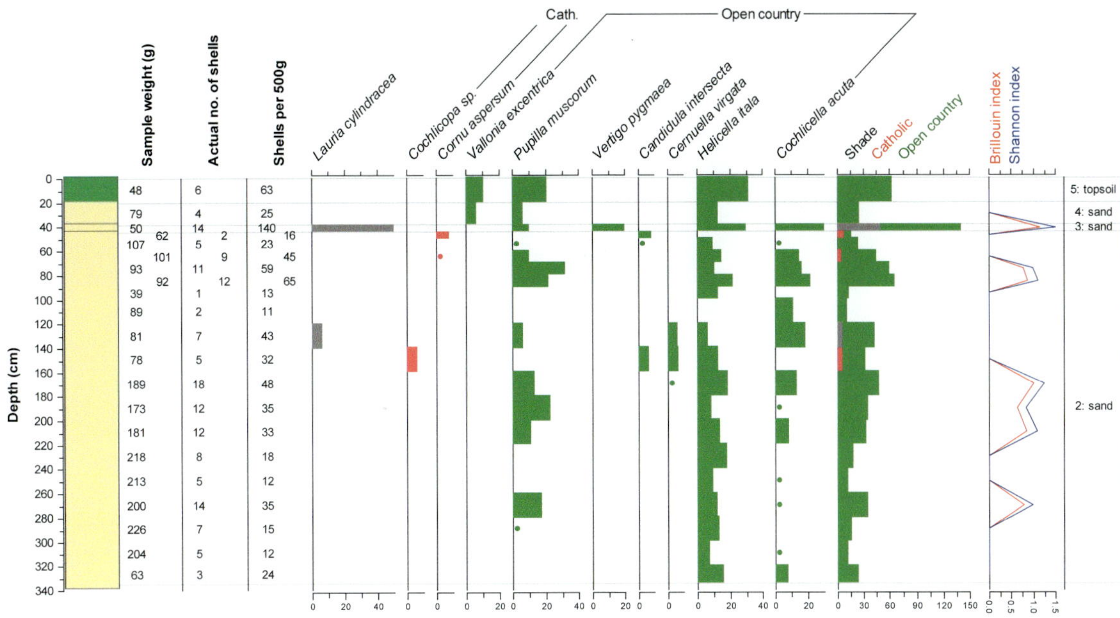

Figure 89. Mollusc diagram for the 289m core.

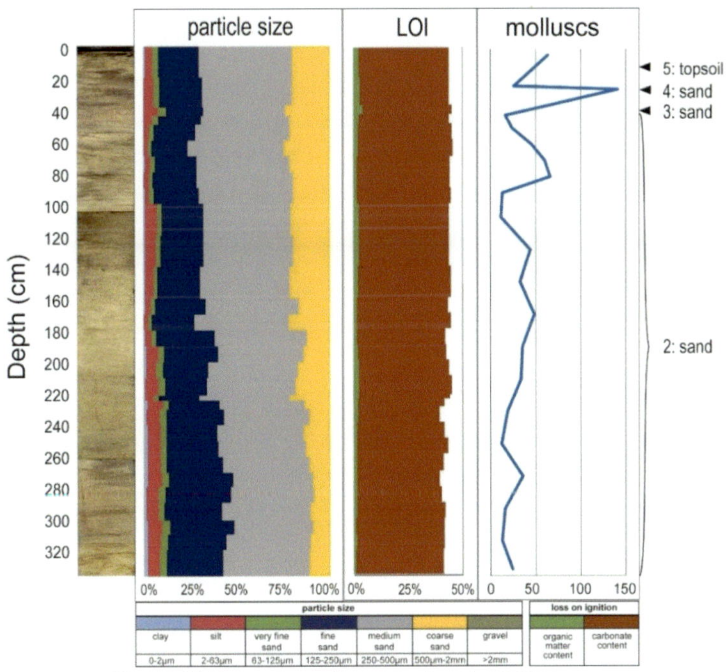

Figure 90. Summary diagrams for the 289m core.

The vegetation on the lower part of the steep slope of the towans changes markedly from that of the meadow. On the slope there is blackthorn (*Prunus avium*), elder (*Sambucas nigra*) and bramble (*Rubus fruticosus*) (Figure 91) and it was difficult to find an adequate sampling location. A position at the upper end of a relatively bare area was selected, just above a more level section of the scarp. The levelled area has been ascribed to sand extraction, referred to in a parish map of 1603 (Fowler and Thomas 1962), and marked on a 1850–51 parish map (Figure 13). The turf sample was poor in molluscs, mostly open country, especially *Cochlicella acuta*, but with some catholic species.

Figure 91. The steep slope of Godrevy Towans showing blackthorn, elder and bramble scrub.
The 289m position is in the centre of the photograph.

The sand was surprisingly deep in this position; the auger could not be driven and extracted below 335cm and it is not known whether this is the level of the bedrock or the sand extends still deeper. The scarp is steep above this location (a gradient of 35% measured from the differential GPS readings between 285m and 320m), and it seems probable that there is a localised hollow in the underlying killas which has allowed this depth of sand to accumulate.

The sand [2] below 44cm shows little variation in particle size, organic matter or carbonate content although there is slightly less silt above 100cm. Mollusc numbers are low with almost entirely open country taxa although occasional shade/catholic species are found. A thin layer of sand at 38–44cm [3] contains a few more shells, particularly with *Lauria cylindracea*, although there are only 5 specimens of this species. Numbers of molluscs are too few to draw any firm conclusions, but it is possible that this was the earliest time when scrub became established on these lower slopes, providing some shade in an otherwise very open and unstable landscape. This layer is overlain by more sand [4] and then topsoil [5], both very sparse in molluscs despite the stabilising effect of the modern scrub growing in the immediate vicinity.

345m core

Grid ref.: SW 58887.42380; modern ground level: 55.90m OD.
Mollusc numbers: able 18; mollusc diagram: Figure 92; summary diagram: Figure 93.

Table 18. Mollusc numbers for the 345m core.

Depth (cm)	0-9	9-15	15-23	23-30	30-40	40-60	60-80	80-105	105-120
Sample weight (g)	284	296	411	316	263	364	372	550	378
Munsell number	10YR 4/2	10YR 5/4	10YR 5/3	10YR 6/4		10YR 7/4			
Zone	**4**	**3**		**2**					**1**
Sediment	topsoil	sand		sand					killas
Carychium minimum									3
Cochlicopa lubrica			1						1
Cochlicopa sp.		1							
Vertigo pygmaea	11	3						1	
Pupilla muscorum			1					2	
Lauria cylindracea	17	5	22	3	1	1		2	
Vallonia excentrica	21		1	1				2	1
Punctum pygmaeum			2						
Vitrina pellucida						1			
Aegopinella nitidula									1
Aegopinella pura		1	4	2					
Balea heydeni		1	4			1			
Cochlicella acuta	76	39	24	4			1	17	
Ashfordia granulata			1						
Candidula intersecta	31	6	2	1					
Cernuella virgata	11	1	2					10	
Helicella itala	14		2			1	3	4	
Xerocrassa geyeri									7
Cepaea sp.	1	1							
Cornu aspersum			1					1	
TOTAL SHELLS	**182**	**58**	**67**	**11**	**1**	**4**	**4**	**39**	**13**
Shells per 500g	320	98	82	17	2	5	5	35	17
TOTAL SPECIES	**8**	**9**	**13**	**5**	**1**	**4**	**2**	**8**	**5**

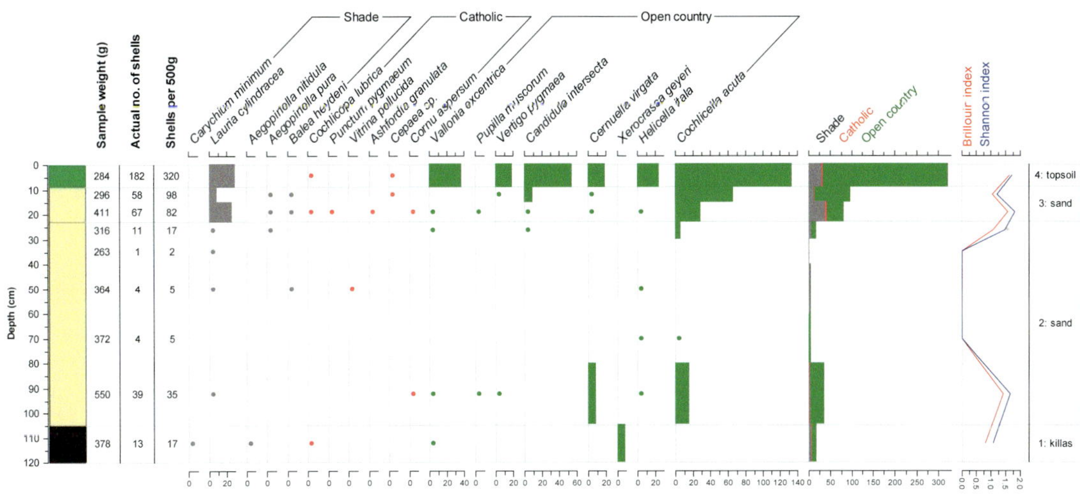

Figure 92. Mollusc diagram for the 345m core.

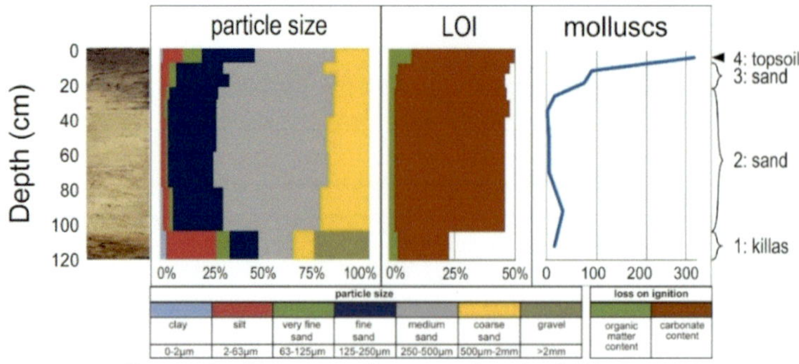

Figure 93. Summary diagrams for the 345m core.

The scrub now growing on the lower part of the scarp fades out about half way up the slope and gives way to very short grazed mixed vegetation (Figure 94) which at 345m includes ragwort (*Senecio jacobaea*), bird's foot trefoil (*Lotus corniculatus*), unidentified moss, heath dog–violet (*Viola canina*) and wild thyme (*Thymus polytrichus*), with very little grass. The modern turf is surprisingly rich in molluscs with moderate diversity (Shannon index 1.78). Nearly all shells are xerophilic taxa (93%) as would be expected from this very exposed area of ground. The presence of a single *Clausilia bidentata* here is anomalous, and it may have been blown in, or brought to the area by birds or other animals.

Figure 94. The hillside around the 345m location, showing very short vegetation on the bare slope.

The killas [1] was reached at a depth of 105cm and contains a few shells, including *Xerocrassa geyeri*, which are presumably derived from the overlying sand. The bedrock is overlain by a thick sand deposit [2] with a moderate number of shells in its lowest level, but few more superficially implying instability for much of the deposition history. The sand from 9–23cm [3] has similar sedimentology to that below but contains much larger numbers of shells with a mixture of shade, catholic and open country species. The presence of *Balea heydeni* (four shells) is very unexpected as this is a geophobic species normally living in woodland and rarely found on the ground, although it is able to live in walls; as with the specimen of *Clausilia* found in the modern turf it is presumably allochthonous. *Lauria cylindracea* is present in good numbers, a taxon well able to live in rock rubble habitats, although perhaps surprising in such numbers on so bare a hillside. The possibility that there were, in the past, areas of woodland or stone walls in this immediate area to account for the *Balea* and *Lauria*, should be considered, but there is no evidence for either at the present time.

The shallow topsoil [4] has large numbers of open country shells indicating that the modern sediment at this location is stable and is not currently subject to major wind deflation or hillwash.

400m core

Grid ref: SW 58886.42435; modern ground level: 66.72m OD.
Mollusc numbers: Table 19; mollusc diagram: Figure 95; summary diagram: Figure 96.

Table 19. Mollusc numbers for the 400m core.

Depth (cm)	0-10	10-22	22-31	31-42	42-50	50-70	70-90	90-100	100-120	120-140	140-160	160-180	180-195	195-211	211-221	221-231	231-241	241-250	250-258	258-268	268-275	275-290
Sample weight (g)	89	55	58	81	70	196	175	47	150	195	162	187	130	69	95	100	105	102	96	104	44	183
Munsel number	10YR 5.3		10YR 3/2	10YR 5/4		10YR 7/4			10YR 5/3	10YR 7/4					10YR 6/6			7.5YR 4/4			10YR 6/4	
Zone	**7**	**6**				**6**									**4**			**3**			**2**	**1**
Sediment	topsoil	buried soil				sand									sand			buried soil			sand	killas
Cochlicopa lubrica				3							3	1						1				
Vertigo pygmaea	2	1		1	1	3			1													
Pupilla muscorum										1								3				
Lauria cylindracea		5	2	5	8	1	3		3	3	7	10	4									
Vallonia excentrica		2	3	1		1				1	1	2						5	1	5	1	
Punctum pygmaeum		1	1	2								4										
Vitrina pellucida		1									1											
Aegopinella nitidula	1			1	1					1		2										
Nesovitrea hammonis		1	1						2													
Cochlicella acuta	5	9	2	10	6		2	3		3	10	18	4		1			11	1	2	1	
Candidula intersecta	1								1						1							
Cernuella virgata			2															2				
Helicella itala	1			4		1				1	1	3	1					4	6	7	1	1
Trochulus hispidus											1											
Cepaea nemoralis									1			1										
Cornu aspersum									1			1						1				
TOTAL LAND	**10**	**20**	**11**	**27**	**16**	**6**	**5**	**3**	**9**	**10**	**24**	**42**	**9**	**0**	**2**	**0**	**0**	**27**	**8**	**14**	**3**	**1**
SHELLS per 500g	56	182	95	167	114	15	14	32	30	26	74	112	35	0	11	0	0	132	42	67	34	3
TOTAL SPECIES	**5**	**7**	**6**	**8**	**4**	**4**	**2**	**1**	**6**	**6**	**7**	**9**	**3**	**0**	**2**	**0**	**0**	**7**	**3**	**3**	**3**	**1**

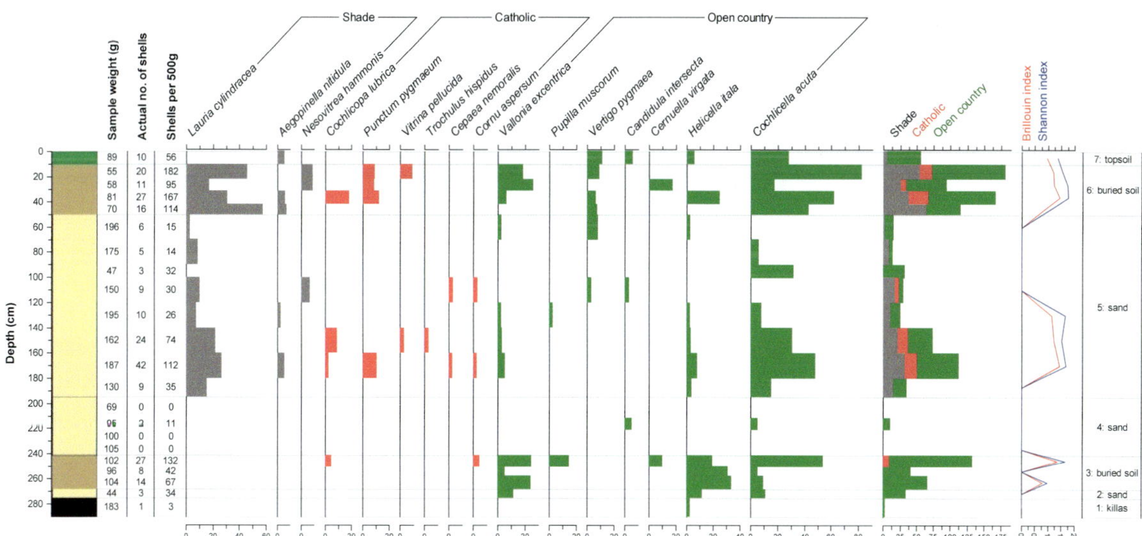

Figure 95. Mollusc diagram for the 400m core.

The 400m location is within an area of tall marram grass (*Ammophila arenaria*) (Figure 97) close to the break of slope at the top of the steepest part of the scarp, although still sloping gently uphill. The density of this grass is reflected in the high proportion of shade species in the turf sample (65%), notably with *Lauria cylindracea* accounting for over one third of all shells. Xerophile species requiring dry exposed habitats are few in number at this site. In particular *Vallonia excentrica* is very scarce, consistent with the greater degree of shade.

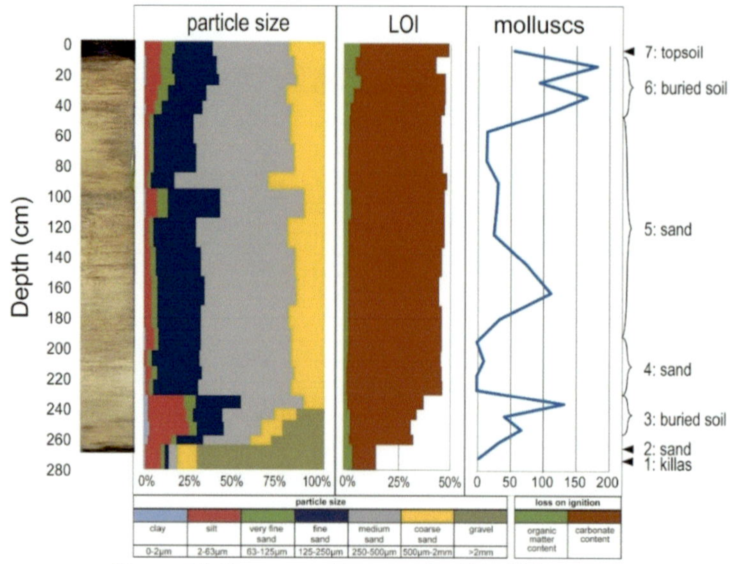

Figure 96. Summary diagrams for the 400m core.

Figure 97. The 400m post is shown within an area of marram grass.

The sand here is deep, the killas [1] being at a depth of 275cm. and it is probable that there is a hollow in the underlying bedrock in which sand had accumulated. A thin sand layer [2] is overlain by a horizon with 25% silt [3] containing a moderate number of molluscs and is interpreted as a buried soil sediment accumulating in the hollow sufficiently slowly for pedogenesis to take place. Unfortunately, the marram grass was too tall to permit a GPR scan of this section of the transect. The presence of *Cernuella virgata* rather than *Xerocrassa geyeri*, together with of *Candidula intersecta* and *Cornu aspersum*, suggests that this deposit is post-Roman and not prehistoric, although bioturbation or winnowing cannot be excluded.

This soil is covered by a 50cm horizon of sand [4] with virtually no shells, and is overlain by 145cm of sand [5] containing varying numbers of molluscs at different depths with *Lauria cylindracea* being prominent, suggesting that there was some shade. Above 50cm silt and organic matter proportions slightly increase and there is also increase in mollusc numbers with a broad spread of habitat requirements, again consistent with buried soil [6] with good vegetation cover, similar to the modern local environment. However, there are few shells in the current topsoil [7] with no *Lauria* despite the modern turf sample from very close to the core position containing numerous *Lauria*. The relative lack of molluscs in the core topsoil (10cm thick) when the modern turf (4–5cm thick) at this site contained large numbers may be ascribed to sampling variation.

451M CORE

Grid ref.: SW 58884.42485; modern ground level: 67.89m OD.
Molluscs numbers: Table 20; mollusc diagram: Figure 98; summary diagram: Figure 99.

Table 20. Mollusc numbers for the 451m core.

Depth (cm)	0-11	11-20	20-32	32-44	44-48
Sample weight (g)	211	399	700	551	206
Munsell number	10YR 3/2	10YR 5/3	10YR 6/4		
Zone	**3**		**2**		**1**
Sediment	topsoil		sand		killas
Cochlicopa lubrica	2	1			
Cochlicopa lubricella	3				
Cochlicopa sp.	2	1	1		
Vertigo pygmaea	12	5			
Pupilla muscorum	1		19		8
Lauria cylindracea	4	2	2		
Vallonia excentrica	78	28	1		3
Punctum pygmaeum	8				
Cochlicella acuta	8	17	41	1	2
Ashfordia granulata			13		
Candidula intersecta	2	4	2		
Cernuella virgata	2	6		1	
Helicella itala		6	4		9
Trochulus hispidus	10	1			
Cepaea sp.	1				
Cornu aspersum			1		
TOTAL SHELLS	**133**	**71**	**84**	**2**	**22**
Shells per 500g	315	89	60	2	53
TOTAL SPECIES	**12**	**9**	**9**	**2**	**4**

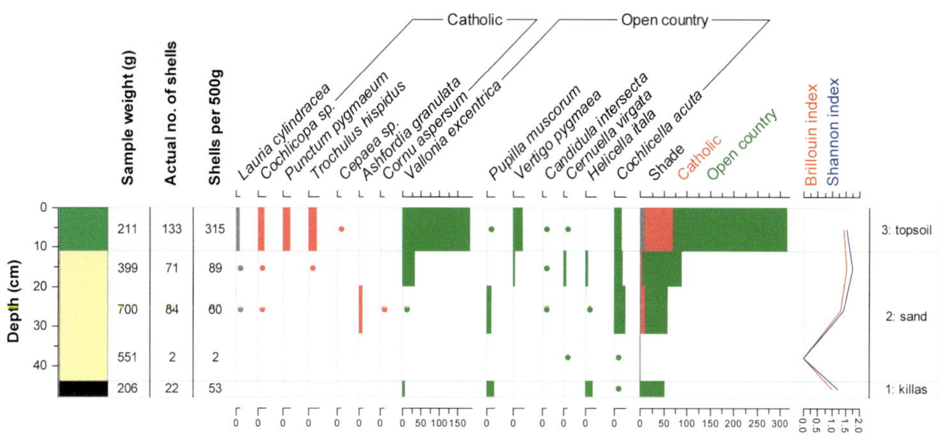

Figure 98. Mollusc diagram for the 451m core.

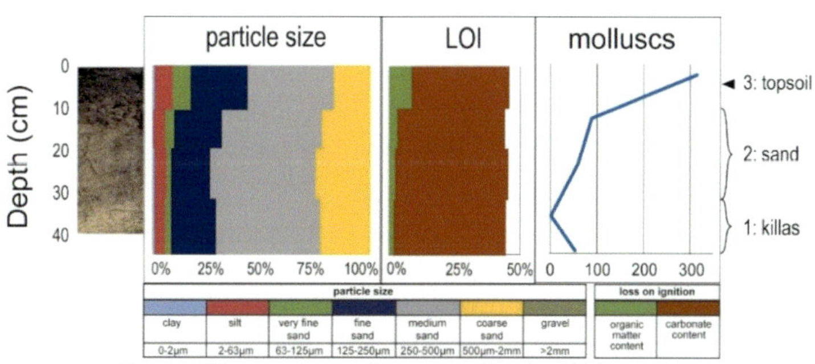

Figure 99. Summary diagrams for the 451m core.

The 451m core is from an area of short grazed grassland about 20m in diameter surrounded by low gorse/bramble scrub and marram grass (Figure 100). There is minimal shade, although hemp-agrimony (*Eupatorium cannabinum*), dandelion (*Taraxacum officinale* agg.) and traveller's joy (*Clematis vitalba*) were found flowering here and provide some shade at their bases. The turf molluscs are variable, with a mixture of species across the habitat range, with *Vallonia excentrica* dominating.

Figure 100. The open short grassland of the 451m position.

The sand is relatively shallow, the killas [1] being at 44cm and containing a few open country molluscs. Above this is a sand layer [2] with a mixed mollusc assemblage with no specific characteristics. The topsoil [3], in contrast, contains a large number of shells, 60% of which are *Vallonia excentrica*, entirely consistent with the short grass grazed by the modern abundant rabbit population. Some shade and catholic species are able to live in the little shade provided by the vegetation.

495m core

Grid ref.: SW 58887.42529; modern ground level: 69.75m OD.
Mollusc numbers: Table 21; mollusc diagram: Figure 101; summary diagram: Figure 102.

Table 21. Mollusc numbers for the 495m core.

Depth (cm)	0-15	15-30	30-38	38-42	42-50	50-65	65-80	80-83
Sample weight (g)	303	675	349	71	279	632	488	123
Munsell number	10YR 3/3			10YR 5/3	10YR 7/4			
Zone	**4**			**3**	**2**			**1**
Sediment	topsoil			sand	sand			killas
Cochlicopa lubrica				1			1	
Cochlicopa sp.	1							
Vertigo pygmaea	7	9		1				
Pupilla muscorum			1	3	6			1
Lauria cylindracea	10	26	11	36	20	2		
Vallonia excentrica	6	4				1		4
Punctum pygmaeum	6	4	1	5	2			
Vitrina pellucida		1	2					
Aegopinella nitidula	3	1	1	4				
Milacidae				1				
Cecilioides acicula	1	3	1			1		
Clausilia bidentata	1	1		3		1		
Cochlicella acuta	12	34	12	8	11	20	4	1
Ashfordia granulata	3	3	1	6				
Candidula intersecta	2	4	2					
Cernuella virgata	1	8	5	3	4	4		3
Helicella itala	6	1	3		4		2	7
Trochulus hispidus	2	1						
Cepaea nemoralis	1	2						
Cepaea sp.	1	1						
Cornu aspersum		1						
TOTAL SHELLS	**63**	**104**	**40**	**71**	**47**	**29**	**7**	**16**
Shells per 500g	104	77	57	500	84	23	7	65
TOTAL SPECIES	**15**	**16**	**11**	**11**	**6**	**6**	**3**	**5**

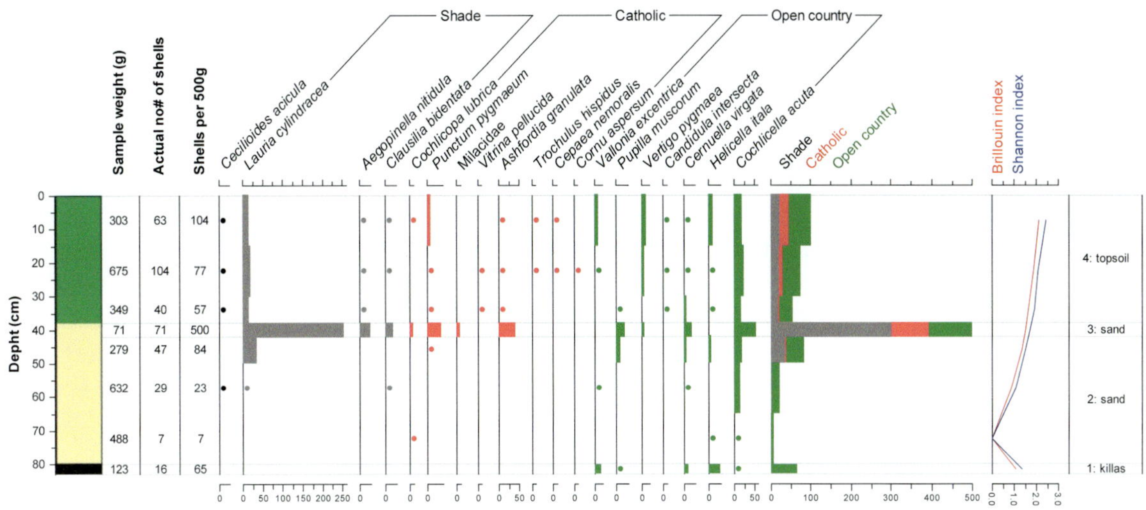

Figure 101. Mollusc diagram for the 495m core.

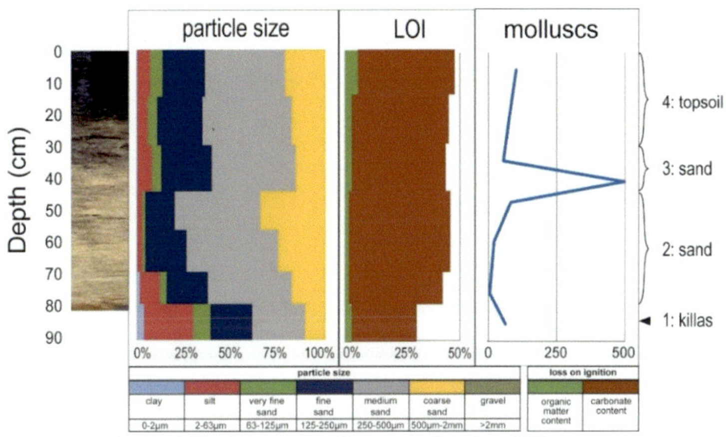

Figure 102. Summary diagrams for the 495m core.

The core position is immediately on the south side of an area of gorse and bramble scrub about 150cm high and is in partial shade but with tall grass (Figure 103). The turf molluscan population shows a large number of taxa (16) with high diversity (Shannon index 2.29). The molluscs include species across the habitat spectrum entirely consistent with its mixed shade/open situation, providing micro-habitats suitable for species requiring both shade and more exposed areas.

Figure 103. Gorse bushes and brambles at the 495m location.

Killas [1] was reached at 80cm, overlain by sand [2] with few molluscs. At 38–42cm there is a thin horizon [3] with numerous *Lauria cylindracea*, as well as *Clausilia bidentata* and *Aegopinella nitidula*, the latter two consistent with the extent of shade provided by the scrub, and therefore not necessarily intrusive. The thick topsoil above this [4] contains shells with mixed environmental requirements but high diversity, very similar to the modern turf block.

Of great interest is the presence of *Cecilioides acicula*, which was also found in the modern turf block. This mollusc is usually ignored in palaeoenvironmental studies as it may burrow to depths of up to 2m, and therefore has little biostratigraphic value. The timing of its introduction to Britain has not been established as there are no reliable prehistoric records (Evans 1972, 168). Its value at the present site is that it is frequently associated with cultivated or disturbed ground (Evans 1972, 168). The location of this core is close to the medieval manor of Crane Godrevy, and about 85m south-west of the closest known extent of the field system to the south of the building (Thomas 1969). The presence of *c. acicula* in this core may, together with the possible finding of a ditch in the 528m core (see below), suggest more extensive agricultural activity than previously thought; the high number of *Lauria cylindracea* supports this conjecture, with the possibility that field walls were, at one time, present in this location to give adequate shade for the *Lauria*.

528m core

Grid ref.: SW 58887.42562; modern ground level: 70.54m OD.
Mollusc numbers: Table 22; mollusc diagram: Figure 104; summary diagram: Figure 105.

Table 22. Mollusc numbers for the 528m core.

Depth (cm)	**0-15**	**15-30**	**30-47**	**47-65**	**65-72**	**72-80**	**80-87**	**87-100**
Sample weight (g)	360	687	785	664	347	374	335	282
Munsell number	10YR 5/3		10YR 7/3	10YR 5/3	10YR 5/6	10YR 7/4	10YR 4/6	
Zone	**4**			**3**			**2**	
Sediment	topsoil			sand			buried soil	
Cochlicopa lubrica		3						
Cochlicopa sp.		5						1
Vertigo pygmaea	14	6				2	1	1
Pupilla muscorum	12	48	1	5	1	4	4	4
Lauria cylindracea		2						
Vallonia excentrica	63	58	5	2		4	16	16
Punctum pygmaeum	2	5			2	1		
Vitrina pellucida		1						
Aegopinella pura	8	2						
Limacidae	1							
Cecilioides acicula	2	1						
Cochlicella acuta	28	173	25	4	8	7	8	6
Candidula intersecta	8	26	2	1				
Cernuella virgata	7	51	3	5	3	5	20	2
Helicella itala	3	76	7		11	12	13	8
Trochulus hispidus	3	5						
Cornu aspersum	1			1				
TOTAL SHELLS	**152**	**462**	**43**	**18**	**25**	**35**	**62**	**38**
Shells per 500g	211	336	27	14	36	47	93	67
TOTAL SPECIES	**13**	**14**	**6**	**6**	**5**	**7**	**6**	**7**
Mytilus sp.								+
Patella sp.							+	+

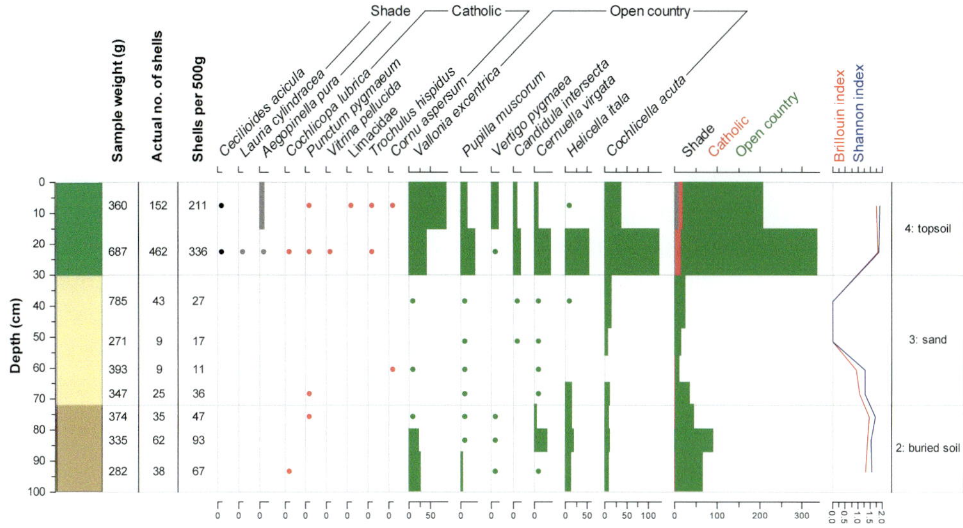

Figure 104. Mollusc diagram for the 528m core.

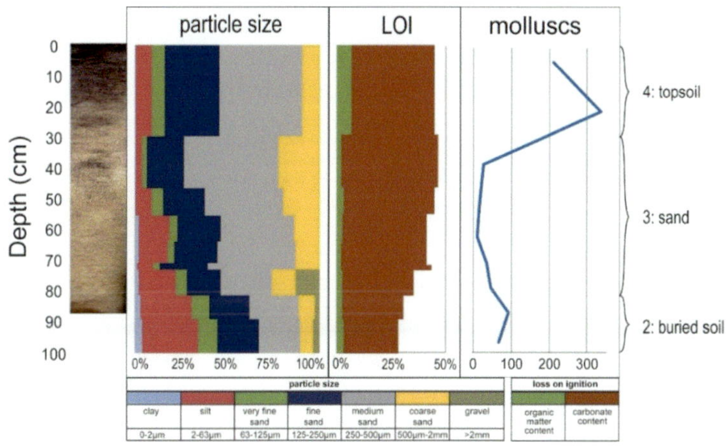

Figure 105. Summary diagrams for the 528m core.

This core was obtained in an area of rank grass at the southern margin of the bare grass of the summit plateau, close to gorse/bramble scrub, and so is largely in shade (Figure 106). This is reflected in the turf molluscs, which include the shade species *Vitrina pellucida* and *Oxychilus alliarius.* as well as good numbers of the open country species *Vallonia excentrica.*

Figure 106. Scrub and rank grass at the 528m core position.

It was only possible to drive the corer to 100cm and it is probable that the bedrock is at this level, but not proven. The lowest sediments [2] contain a high proportion of clay/silt (up to 30%) with a moderate number of molluscs, nearly all open country taxa, and is consistent with a buried soil. This is overlain by sand [3] with fewer shells, again mostly open country species. The topsoil [4] is rich in molluscs, some of which are shade and catholic species consistent with its shady location.

The finding of *Cernuella virgata* and *Candidula intersecta* in the base levels with *Cornu aspersum* in the intermediate sand and *Cecilioides acicula* in the topsoil implies that this is disturbed sand throughout its depth. The ground penetrating radar (Figure 107) shows a dip in the bedrock profile at 530–535m, very close to the position of this core. The combination of the radar and mollusc findings suggest that there may have been a ditch in this area of the towans, perhaps defining a field system, and with disturbance of the sands during medieval times associated with the nearby Godrevy manor house. This is approximately 50m south-west of the nearest ditch revealed in previous excavations (Thomas 1969; Sturgess and Lawson Jones 2006b, Figure 2) and, together with the findings described in the 495m core, may indicate a previously unknown area of activity. The presence of the marine mussels (*Mytilus* sp.) and limpets (*Patella* sp.) suggests the use of seaweed as manure in these fields.

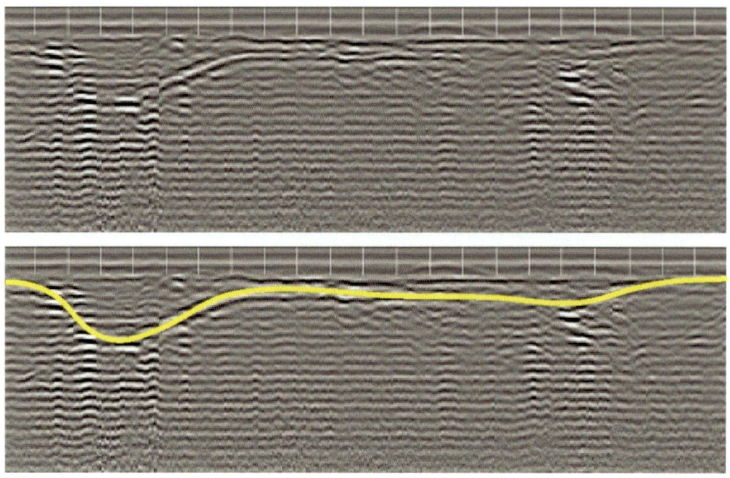

Figure 107. Ground penetrating radar (400MHz) from 528m to 550m.
The depression in the line of the bedrock from 530–535m is drawn on the lower trace.

551m CORE

Grid ref.: SW 58887.42585; modern ground level: 70.26m OD.
Mollusc numbers: Table 23; mollusc diagram: Figure 108; summary diagram: Figure 109.

Table 23. Mollusc numbers for the 551m core.

Depth (cm)	0-9	9-18	18-37	37-50
Sample weight (g)	215	514	843	564
Munsell number	10YR 3/3		10YR 7/4	
Zone	**3**		**2**	**1**
Sediment	topsoil		sand	killas
Vertigo pygmaea	10	17		
Pupilla muscorum	14	70	4	2
Lauria cylindracea		2		
Vallonia excentrica	49	39	2	4
Punctum pygmaeum	10	12	1	
Aegopinella nitidula		4		
Cochlicella acuta	7	105	12	10
Ashfordia granulata		3		
Candidula intersecta	4	31	2	
Cernuella virgata	4	48	5	9
Helicella itala	1	45	7	5
Cornu aspersum		1		
TOTAL SHELLS	**99**	**377**	**33**	**30**
Shells per 500g	230	367	20	27
TOTAL SPECIES	**8**	**12**	**7**	**5**

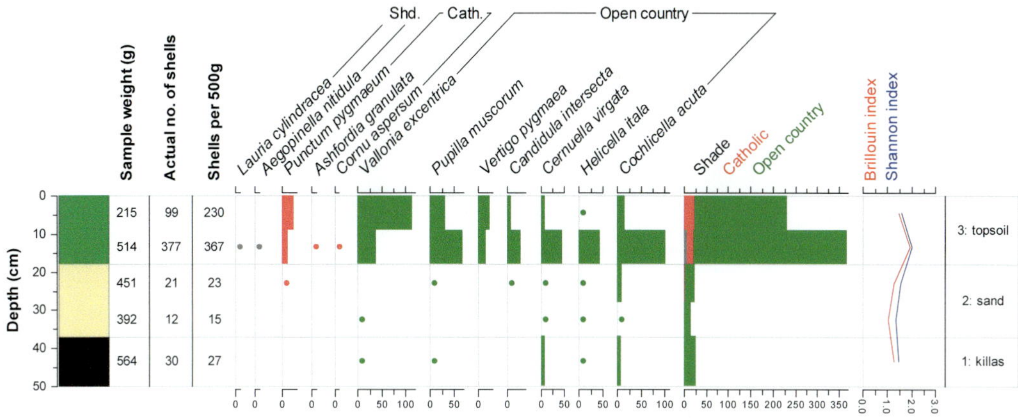

Figure 108. Mollusc diagram for the 551m core.

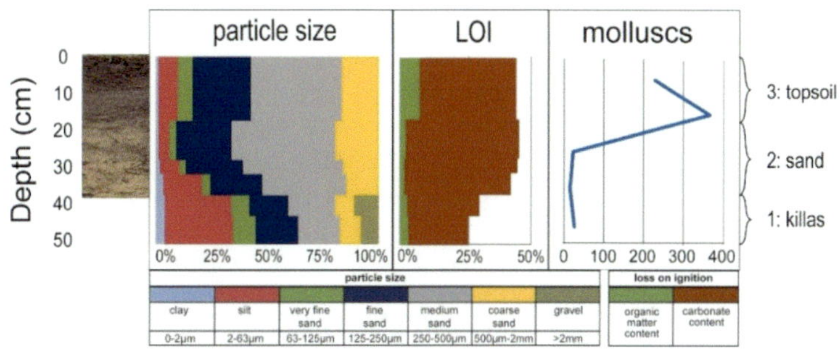

Figure 109. Summary diagrams for the 551m core.

The summit of Godrevy Towans is a level plateau with large areas of short grazed grass and with gorse/bramble patches of scrub to the north and south (Figure 110), but open to the east and west, and very exposed to storms blowing in from the Atlantic Ocean to the west, there being no higher ground or vegetation between this location and the coast. Wild thyme (*Thymus polytrichus*), columbine (*Aquilegia vulgaris*) and sheep's fescue (*Festuca ovina*) were all growing in the area.

Figure 110. The flat grazed summit of Godrevy Towans, looking northward.
The fence at the west edge of the Crane Godrevy manor house site is on the right of the photograph.

The core is from the centre of this bare grassland. The sand is shallow, with the bedrock [1] at 39cm and containing a few open country molluscs. The blown sand [2] over this is relatively sterile as far as molluscs are concerned but there are good numbers of shells in the topsoil [3], mostly open country in keeping with the local environment, but with some *Punctum pygmaeum*, a species capable of living in a very wide variety of habitats, including open, dry grassland. Romano-British or later introductions (*Cernuella virgata*, *Candidula intersecta*) are present at all levels with a single *Cornu aspersum*, either due to bioturbation or regular removal and replacement of the sands by storms blowing over the summit of the towans.

570m core

Grid ref.: SW 58887.42604; modern ground level: 70.56m OD.
Mollusc numbers: Table 24; mollusc diagram: Figure 111; summary diagram: Figure 112.

Table 24. Mollusc numbers for the 570m core.

Depth (cm)	0-17	17-30
Sample weight (g)	696	532
Munsell number	10YR 4/2	10YR 5/4
Zone	**3**	**2**
Sediment	topsoil	sand
Vertigo pygmaea	7	2
Pupilla muscorum	22	3
Vallonia excentrica	29	22
Punctum pygmaeum	4	
Aegopinella nitidula	1	
Aegipinella pura	1	
Cochlicella acuta	19	15
Candidula intersecta	23	
Cernuella virgata	6	21
Helicella itala	5	15
Cornu aspersum		1
TOTAL SHELLS	**117**	**79**
Shells per 500g	84	74
TOTAL SPECIES	**10**	**7**
Mytilus sp.		+

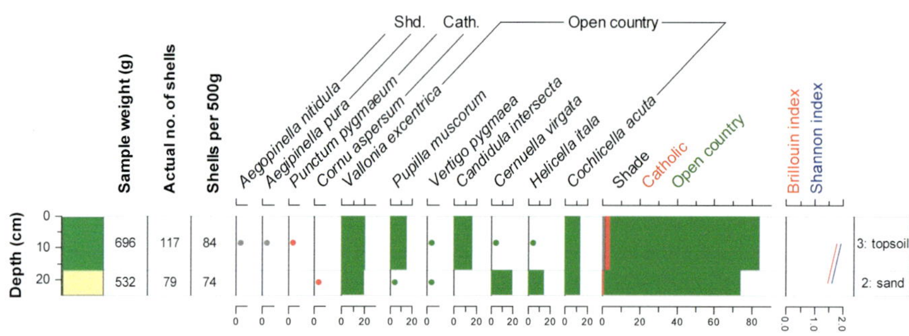

Figure 111. Mollusc diagram for the 570m core.

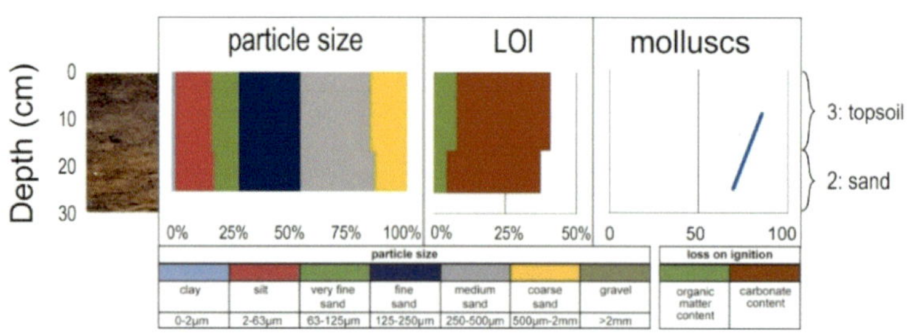

Figure 112. Summary diagrams for the 570m core.

The sand in this core, also from the open grazed grassland exposed to the weather from the west, is very shallow, with bedrock at only 30cm. There is an overlying thin layer of sand [2] with a covering of topsoil [3]. Molluscs, mainly open country species, are present in moderate numbers with the variety consistent with the habitat, and are similar to the modern turf molluscs, although with fewer *Vertigo pygmaea* and *Helicella itala*. The shade species *Aegopinella nitidula* and *A. pura* are surprising; they may well have blown here from the surrounding scrub but the base of the short vegetation may provide sufficient shade necessary for these species to live. A mussel fragment at the lower level may have been dropped by birds or come from the nearby settlement at Crane Godrevy.

582M CORE

Grid ref.: SW 58887.42616; modern ground level: 70.39m OD.
Mollusc numbers: Table 25; mollusc diagram: Figure 113; summary diagram: Figure 114.

Table 25. Mollusc numbers for the 582m core.

Depth (cm)	0-15	15-28	28-35	35-43
Sample weight (g)	326	626	318	57
Munsell number	10YR 4/2		10YR 5/3	
Zone	3	2		1
Sediment	topsoil	sand		killas
Vertigo pygmaea	14	14		
Pupilla muscorum	11	25	2	
Vallonia excentrica	80	28	6	
Punctum pygmaeum	18	8		
Aegopinella nitidula	4			
Aegopinella pura	2			
Limacidae		1		
Cochlicella acuta	4	51	23	
Candidula intersecta	7	9		
Cernuella virgata		17	5	
Helicella itala		38	15	
Trochulus hispidus		1		
TOTAL SHELLS	**140**	**192**	**51**	**0**
Shells per 500g	215	153	80	
TOTAL SPECIES	**8**	**10**	**5**	
Mytilus sp.		+	+	

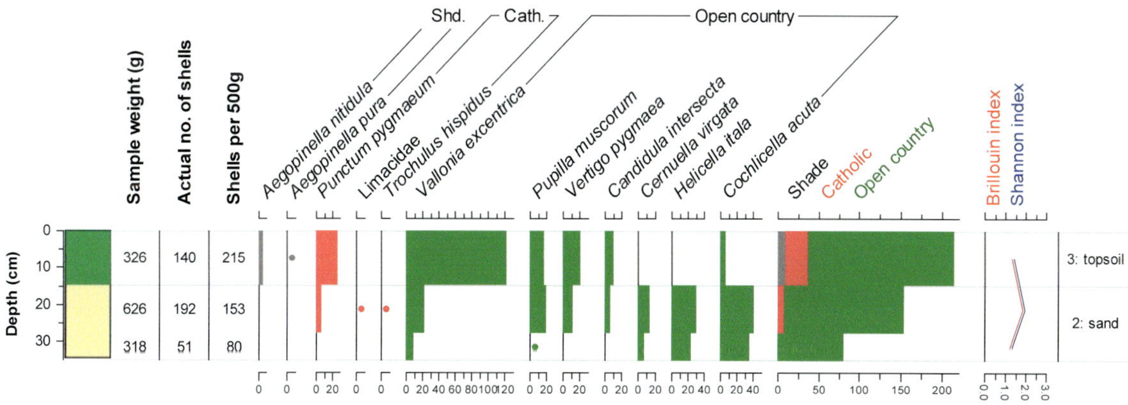

Figure 113. Mollusc diagram for the 582m core.

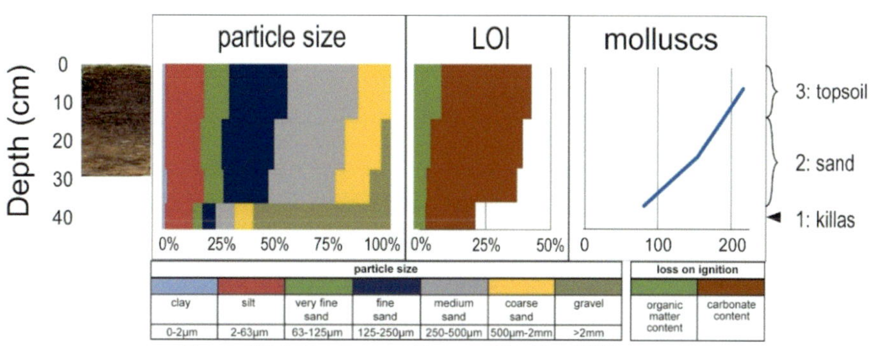

Figure 114. Summary diagram for the 582m core.

The final core in the transect is from the north edge of the open summit grassland, in an area surrounded on three sides with 2m high gorse (Figure 115), with turf molluscs strongly dominated by *Vallonia excentrica* which accounts for over 70% of shells present. As in the 570m core the sediments are shallow with the bedrock [1] at 35cm. The sand above this [2] contains open country taxa with a few catholic species while the topsoil [3] has a high number of shells, again mostly open country species, but more catholic shells, including *Aegopinella nitidula* and *A. pura* which are more likely to be autochthonous here than at the 570m site.

Figure 115. The 582m site with gorse encircling on three sides.

Chapter VIII – Pollen analysis at Gwithian.

by Dr C. R. Batchelor, Quaternary Scientific, University of Reading

INTRODUCTION

This report summarises the findings arising out of the pollen analysis undertaken on samples taken from selected horizons of the 0.5m Gwithian core. The overarching aim of the pollen analysis was to reconstruct the vegetation history of the site and its environs, and to identify any evidence of human activity. The dates quoted in this chapter are derived from those extrapolated from the OSL and radiocarbon dates determined on the 30m core, and explained in the following chapter.

METHODS

Pollen assessment and analysis

Twenty pollen samples were extracted as follows: (1) sampling a standard volume of air-dried sediment (1cm^3); (2) addition of two *Lycopodium* tablets to enable calculation of pollen concentrations; (3) deflocculation of the sample in 1% sodium pyrophosphate; (4) sieving of the sample to remove coarse mineral and organic fractions (>125μm); (5) acetolysis; (6) removal of finer minerogenic fraction using sodium polytungstate (specific gravity of 2.0g/cm^3); (7) mounting of the sample in glycerol jelly stained with safranin. Each stage of the procedure was preceded and followed by thorough sample cleaning in filtered distilled water. Quality control was maintained by periodic checking of residues, and assembling sample batches from various depths to test for systematic laboratory effects. Pollen grains and spores were identified using the University of Reading pollen type collection and keys and photographs (Moore *et al.* 1991; Reille 1992). Plant nomenclature follows the Flora Europaea as summarised in Stace (2005). The assessment procedure consisted of scanning the prepared slides, and recording the concentration and preservation of pollen grains and spores, and the principle taxa on four transects (10% of the slide). The analysis procedure consisted of counting the prepared slides to a minimum of 300 terrestrial land pollen species per level. Pollen percentages are calculated based on terrestrial plants. Aquatic and fern spores are calculated as a percentage of terrestrial pollen plus the sum of the component taxa within the respective category. The number of microcharcoal fragments >40um on at least one axis was also recorded and calculated as a concentration per cm^3. The results are displayed in Figure 1 together with the geoarchaeological and chronological datasets and inferred environmental reconstruction.

RESULTS, INTERPRETATION AND DISCUSSION

The results of pollen analysis are presented as a percentage pollen diagram (Figure 116). The age/depth model has been obtained from the OSL and radiocarbon dates shown in the left column of the diagram. Pollen concentration and preservation was high in the majority of samples, and a count of 300 terrestrial pollen grains was achieved in 17 of the 20 samples. The pollen diagram was divided into three local pollen assemblage zones (LPAZ's) as shown in the figures.

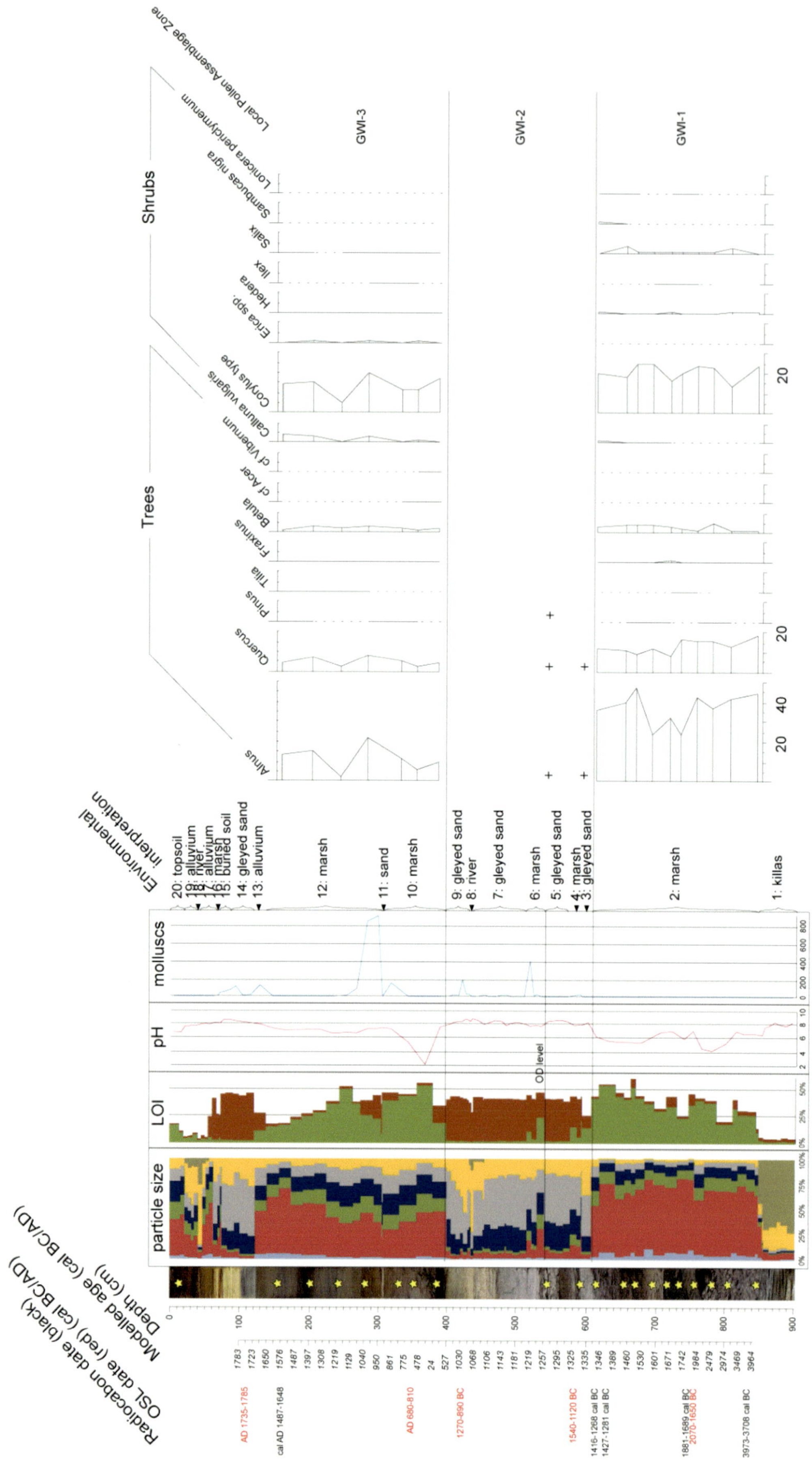

Figure 116. Percentage pollen diagram from Gwithian, Cornwall, also incorporating the main geoarchaeological and chronological datasets and inferred environmental reconstruction.

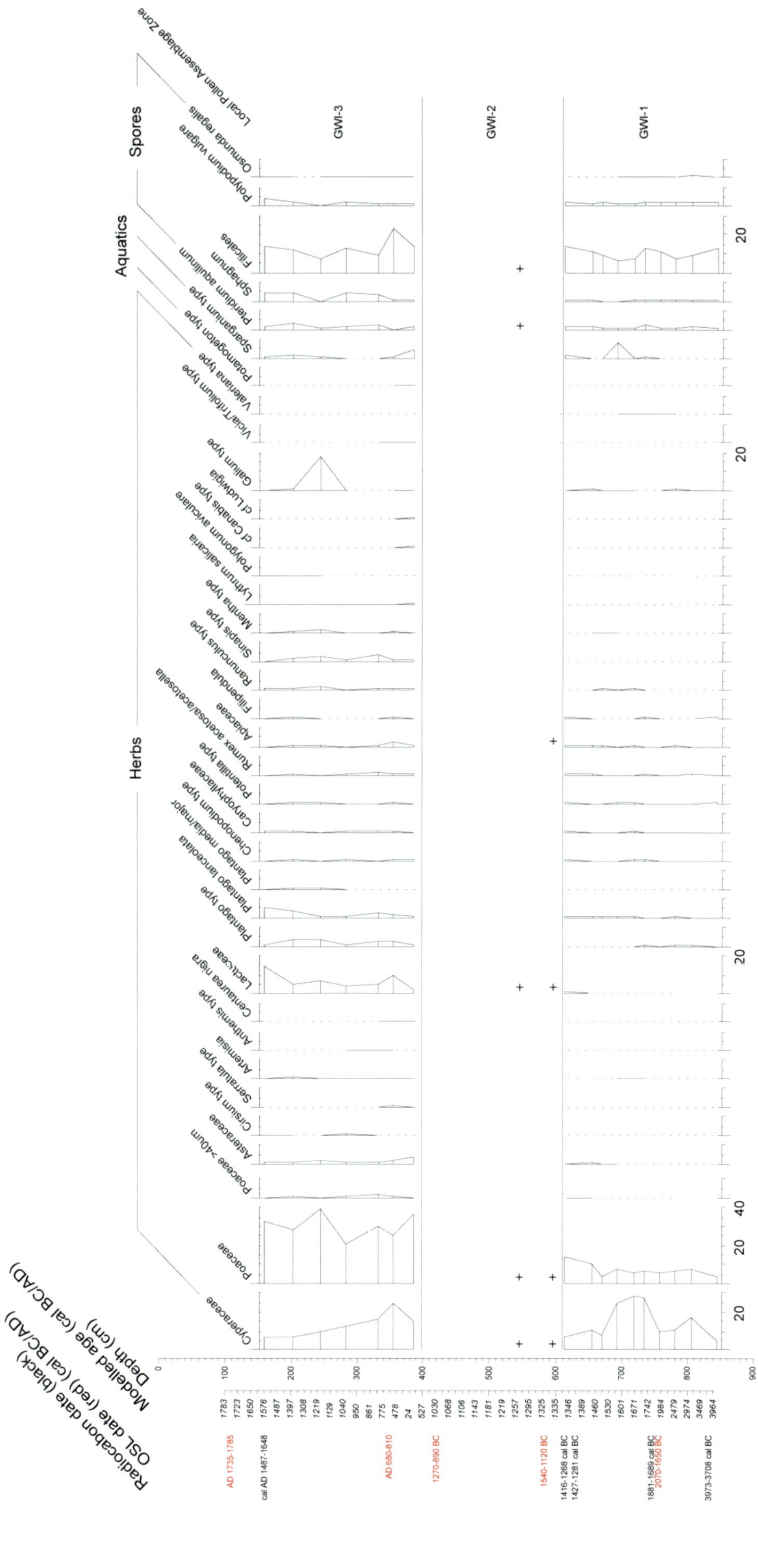

Figure 116 (continued).

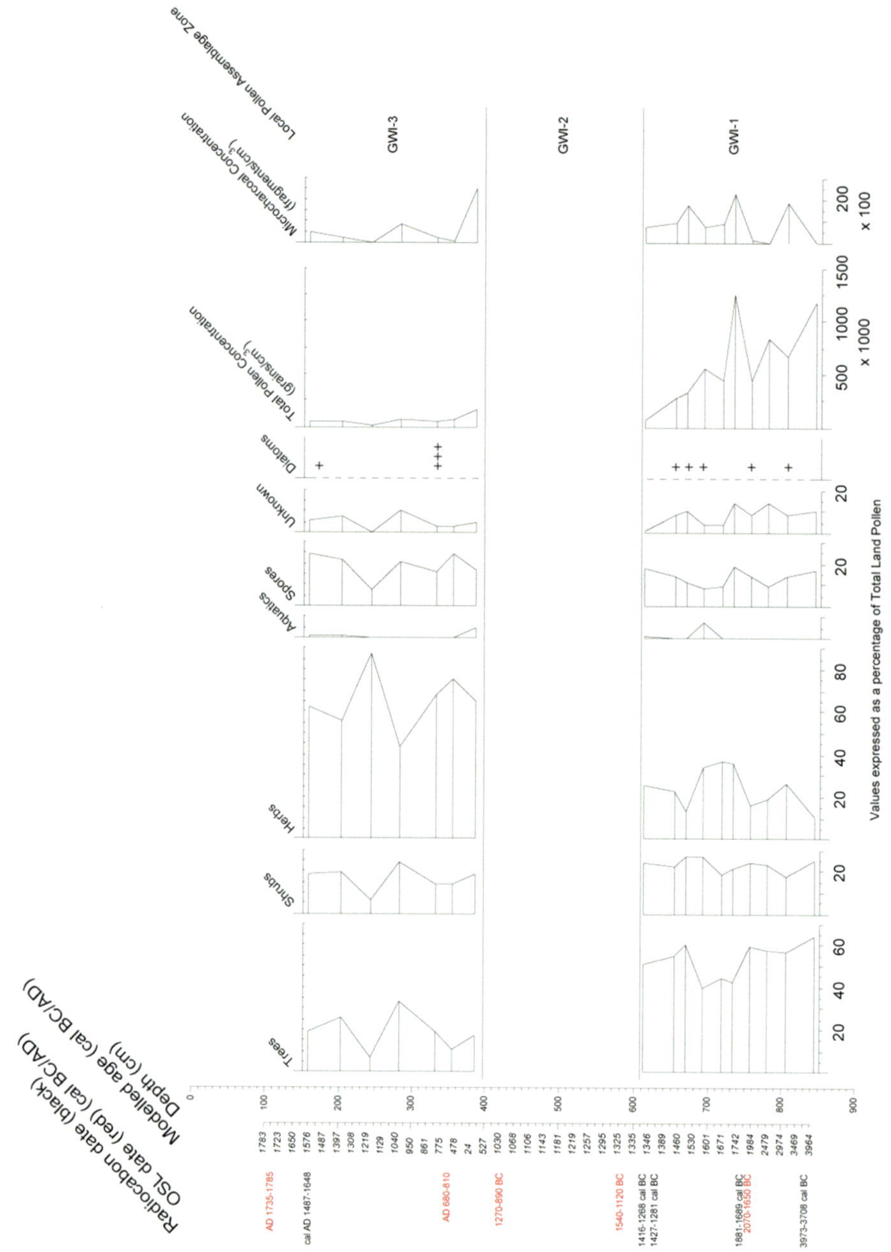

Figure 116 (continued).

LPAZ GWI–1 ***8.46 to 6.14m BGL*** ***Alnus – Corylus type – Quercus***

3965 **to** ***1340*** **cal BC**

Ten samples comprise LPAZ GWI–1, all of which are characterised by high values of tree (60%) and shrub (20%) pollen. *Alnus* dominates (40%) with *Corylus* type (20%), *Quercus* (15%), *Betula*, *Salix* (<5%) and sporadic occurrences of *Pinus*, *Tilia*, *Fraxinus*, *Calluna vulgaris*, *Hedera*, *Ilex*, *Sambucas nigra* and *Lonicera periclymenum* (all <3%). Herb taxa (20–30%) are dominated by Cyperaceae (15–30%) with Poaceae (5%), *Plantago lanceolata*, *Chenopodium* type, Apiaceae and *Galium* type (all <3%). Aquatic taxa comprise moderate values of *Sparganium* type only (<5%). Spore values are dominated by *Filicales* (10%) with *Pteridium aquilinum*, *Sphagnum* and *Polypodium vulgare* (all <5%). Total pollen concentration is high, but declines through the zone. Microcharcoal is present throughout.

LPAZ GWI–2 ***6.14 to 3.86m BGL*** ***Minimal pollen***

1340 **to** ***25*** **cal BC**

Two samples comprise LPAZ GWI–2, both of which are characterised by a very poor concentration of pollen in a very poor state of preservation. Only sporadic occurrences of *Alnus*, *Quercus*, *Pinus*, Cyperaceae, Poaceae, Lactuceae, Apiaceae, *Filicales* and *Pteridium aquilinum* were recorded.

LPAZ GWI-3 ***3.86 to 1.57m BGL*** ***Poaceae – Corylus type – Alnus***

***25* cal BC to *1550* cal AD**

Seven samples comprise LPAZ GWI-3, all of which are characterised by high values of herbaceous pollen (70%). Poaceae dominates (30%) with Cyperaceae (10–15%), Lactuceae, *Plantago* type, *Plantago lanceolata*, *Ranunculus* type, *Rumex acetosa/acetosella*, Apiaceae, *Sinapis* type, *Chenopodium* type, Caryophyllaceae and Poaceae >40um. Tree (15–20%) and shrub (15–20%) taxa are dominated by *Corylus* type (15%) and *Alnus* (10%) with *Quercus* (5%), *Betula*, *Calluna vulgaris* and *Erica* spp. (all <5%). Aquatic taxa (<5%) include *Sparganium* type with sporadic values of *Potamogeton* type. Spores are dominated by *Filicales* (15%) with *Sphagnum* (5%), *Pteridium aquilinum* and *Polypodium vulgare* (both <5%). Total pollen concentration is generally low. Microcharcoal is present throughout.

Interpretation and discussion of the Gwithian pollen analysis

Local Pollen Assemblage Zone GWI-1

The results of the pollen-stratigraphical analysis indicate that during LPAZ GWI-1 (3964–1340 cal BC; 8.46–6.14m below ground level), *Alnus* (*Alnus glutinosa* – common alder) and *Salix* (willow) occupied a large part of the wetland environment forming fen/carr woodland, probably with sporadic *Hedera* (ivy) and elder (*Sambucas nigra*). The presence of a range of herbaceous and aquatic taxa including sedges (Cyperaceae), grasses (Poaceae, probably including common reeds (*Phragmites australis*)), bur-reed (buttercups/water crowfoots (*Ranunculus* type), marsh valerian (*Valeriana* type), mint (*Mentha* type) and ferns (*Polypodium vulgare* / *Filicales*) suggests a damp floodplain surface with areas of still or slowly moving water. There is little indication of changing conditions, with the exception of an apparent shift towards wetter conditions between 1742–1460 cal BC (740–660cm), as indicated by a decline in alder/willow and increase of sedges/bur-reed. No definitive taxa are recorded that might indicate saline conditions (e.g. *Armeria* type – thrift). Values of *Chenopodium* type and *Plantago* type pollen might be indicative of nearby salt marsh plants such as annual seablite (annual seablite) and sea plantain (*Plantago maritime*). However, Chenopodiaceae and *Plantago* plants can also occupy dryland environments indicating disturbed ground (e.g. *Chenopodium album* – fat hen) or open grassland (e.g. *Plantago lanceolata* – ribwort plantain).

Hazel (*Corylus* type) and birch (*Betula*), may also have occupied the floodplain environment, but are more likely to have grown on the dryland, forming mixed deciduous woodland/shrubland with oak (*Quercus*). The limited quantity of oak and relatively high values of hazel indicate that this woodland was relatively open, and perhaps limited in extent. Elm (*Ulmus*) does not appear to have formed part of the dryland woodland, either suggesting it did not naturally grow in this area, or that the sequence post-dates the early Neolithic elm decline (*c.* 4400–3340 cal BC) (Parker *et al.* 2002; Batchelor *et al.* 2014). Similarly, lime (*Tilia*) either did not naturally occupy the area, or had already undergone a decline; the decline of lime is normally recorded across the British Isles during the late Neolithic/early Bronze Age, often with other woodland taxa (e.g. Grant *et al.*, 2012).

Quercus values decline from midway through the zone (*c.* 1671 cal BC; 7.20m BGL), suggestive of a reduction in dryland woodland. This coincides with the increase of sedges/bur-reed outlined above, and an apparent change in sedimentology, possibly indicating the decline is linked to changes taking place on the floodplain (e.g. paludification: Waller 1994; Grant *et al.* 2011). However, whilst definitive evidence of human activity is lacking (e.g. cereal pollen), anthropogenic interference is considered the most likely cause of the decline due to: (1) the timing of the event (early Bronze Age); (2) the near contemporaneous increase of herbaceous taxa (some of which may be suggestive of disturbed ground (e.g. *Chenopodium* type (e.g. fat hen) and *Plantago lanceolata* (ribwort plantain)) and (3) relatively high vales of microcharcoal. This interpretation is enhanced by nearby archaeological excavations, which revealed evidence of Bronze Age settlement and farming (Nowakowski *et al.* 2006). Also of note is the occurrence of heather (*Calluna vulgaris*) pollen in low values following the decline of oak, which may be indicative of the development of nearby acidic soils and heathland. This may even represent a coastal heath community including taxa such as ribwort plantain and cinquefoil (*Potentilla* type).

Local Pollen Assemblage Zone GWI–2

This zone (*1340–24* cal BC; 6.14 to 3.86m BGL) is characterised by very low concentrations of pollen in a very poor state of preservation. These values are most likely a reflection of the dominantly sandy sediments deposited during this period causing physical deterioration of the grains.

Local Pollen Assemblage Zone GWI–3

The pollen assemblage from GWI–3 is indicative of a reduced woodland cover on both the wetland and dryland. On the wetland, a more open and wetter environment is indicated by the reduction of alder and willow, and increase of herbaceous and aquatic taxa including grasses (probably including reeds) and sedges together with bur-reed, mint, buttercup/water crowsfoot and loosestrife (*Lythrum salicaria*). Similar to LPAZ GWI–1, no definitive indicators of a saline influence are recorded.

On the dryland, some woodland/shrubland dominated by hazel and oak is indicated, but this is relatively open and much reduced to LPAZ GWI–1. The high quantity of grasses together with ribwort plantain, dandelion (Lactuceae), dock/sorrel (*Rumex acetosa/acetosella*), daisies (Asteraceae), plants from the Brassica family (*Sinapis* type) and possibly cereal pollen (Poaceae >40um) is suggestive of an anthropogenically modified dryland landscape, including suggestions of disturbed land, grassland, crops and their associated weeds. LPAZ GWI–3 is also characterised by higher values of heather (*Calluna vulgaris*) pollen, *Sphagnum* (moss) spores, and the occurrence of Erica spp. (heath). This is suggestive of an increase in the size or proximity of the acidic soils and heathland. As outlined in LPAZ GWI–1, this could be representative of a coastal heath community with cinquefoil and ribwort plantain.

Chapter IX - Chronology and discussion of the coring transect

Dates have been obtained at several locations at Gwithian, but only those from the 30m core will be discussed in this chapter, with extrapolation to the 0.5m core to permit correlation with the pollen and geochemistry studies. The dates obtained from the trench samples and at Strap Rocks will be covered in the relevant chapters. All dates are given as calendar years BC/AD; this overcomes the problem that radiocarbon dates by convention use 1950 as the 'present' date whereas OSL uses the year of measurement or a stated datum close to the year of measurement. Also, radiocarbon dates ned to be presented as calibrated dates while OSL measures time before the year of dating. Radiocarbon dates have been dated using OxCal version 4.3 and IntCal 13 Curves (Reimer *et al.* 2013).

Periods have been defined as Mesolithic: to 4000 cal BC; Neolithic: 4000–2200 cal BC; early Bronze Age: 2200–1500 cal BC; middle Bronze Age: 1500–1150 cal BC; late Bronze Age: 1150–750 cal BC; Iron Age: 750 cal BC–cal AD 50; Romano-British and medieval: cal AD 50–1500; recent: cal AD 1500–present (Needham 1996; Needham *et al.* 1997; Needham 2012).

The 30m core was selected for scientific dating as the sediments within the core tubes had not been disturbed other than splitting and were therefore suitable for OSL dating. The sediments contained several sand deposits suitable for OSL as well as organic sediments for radiocarbon dating. Selection of sample levels for OSL was determined by the presence of sandy deposits to provide a spread of dates along the length of the core. Selection of material for radiocarbon dating depended on the presence of organic material from the marsh levels, with molluscs being used for intermediate levels. Radiocarbon dating was obtained from a NERC Radiocarbon Facility award (grant no. NF/2013/1/14) and performed by the Oxford Radiocarbon Accelerator Unit. OSL dating was performed at ALRL, Aberystwyth University. Identification of wood and charcoal samples was performed by Phil Austin (University College London); Dan Young (University of Reading) identified the seeds.

The dates obtained are shown in Table 26 (OSL dates) and Table 27 (radiocarbon dates). Figure 117 is an age/depth plot, with trend lines linking all except the 384cm shell radiocarbon date which appears anomalous. With that exception, all of the dates fall in chronological sequence (allowing for 2σ (radiocarbon) or 1σ (OSL) variation). The graph shows alternating phases of slow and rapid deposition of sediments - slow to the early Bronze Age, rapid through the Bronze Age, slow to the mid post-Roman/medieval period, and then rapid again to the present time.

Table 26. OSL dates on the 30m core samples.

Lab number	Corrected depth (cm)	OSL date (1σ) (years before 2010)	Calendar date (1σ)	Mid-date
Aber-203/GT12-B30-1	150 ± 3	250 ± 25	AD 1735–1785	AD 1760
Aber-203/GT12-B30-2	384 ± 3	1290 ± 80	AD 640–800	AD 770
Aber-203/GT12-B30-3	463 ± 3	3060 ± 240	1290–810 BC	1050 BC
Aber-203/GT12-B30-4	665 ± 1	3350 ± 210	1550–1130 BC	1340 BC
Aber-203/GT12-B30-5	903 ± 2	3880 ± 210	2080–1660 BC	1870 BC

Table 27. Radiocarbon dates on the 30m core samples.

Lab number	Material	Corrected depth (cm)	$\delta^{13}C$	Radiocarbon date (BP)	Calibrated date (2σ)	Mid-date
OxA-29165	*Menyanthus trifoliata* seeds	180–207	-25.97‰	313 ± 29	cal AD 1485–1645	AD 1565
OxA-28960	*Cochlicella acuta*	381–387	-5.98‰	1975 ± 26	45 cal BC–cal AD 75	AD 15
OxA-28959	*Cochlicella acuta*	491–506	-7.50‰	2837 ± 28	1090–910 cal BC	1000 BC
OxA-28958	cf. *Alnus glutinosa* wood	705–718	-28.75‰	3081 ± 27	1420–1265 cal BC	1340 BC
OxA-28957	*Alnus glutinosa* wood	718–740	-27.15‰	3093 ± 29	1430–1280 cal BC	1355 BC
OxA-29164	cf. *Alnus glutinosa* charcoal	835–867	-28.06‰	3455 ± 34	1885–1685 cal BC	1785 BC
OxA-28956	*Alnus glutinosa* wood	949–977	-27.90‰	5016 ± 32	3945–3705 cal BC	3825 BC

(an additional sample, at 777–804cm, was submitted but failed to produce a date due to excessively low yields).

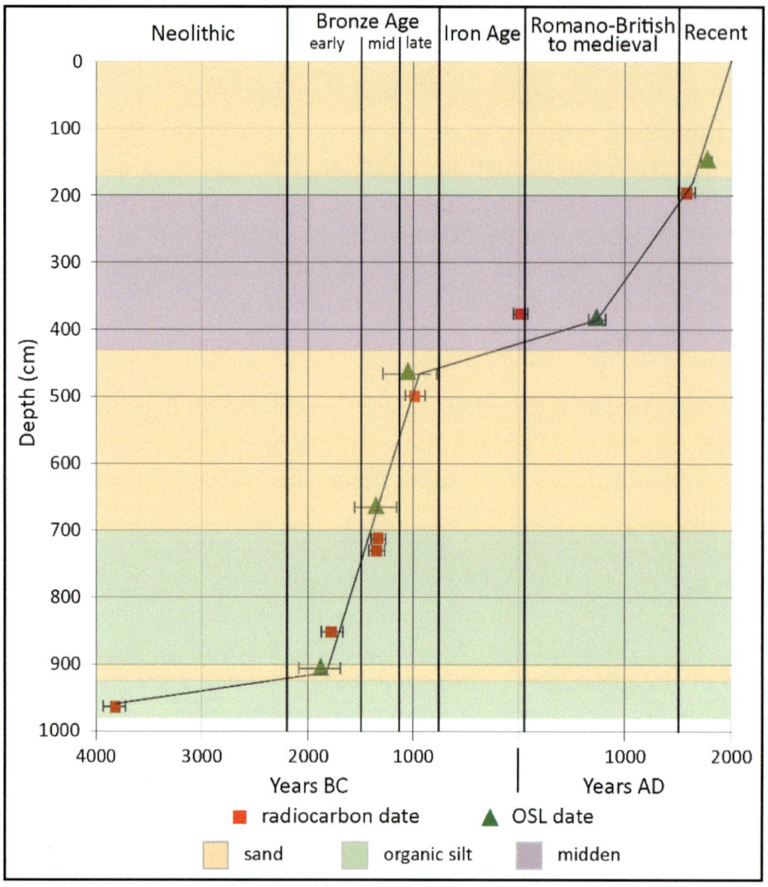

Figure 117. Time/depth plot for the 30m core.

The organic silts of the early marsh formed over at least 2000 years, from *c.* 4000–*c.* 2000 BC, but after an episode of sand blow there was much more rapid accumulation of silt up to the middle of the mid Bronze Age. A thick deposit of sand built up on the marsh at the same overall rate from the middle of the mid Bronze Age to about the middle of the late Bronze Age, but it cannot be determined whether this was constant or if there were varying episodes of rapid and slow sand deposition, with or without periods of deflation. The dating of the midden will be discussed below, but it is unlikely that the graph accurately reflects the time when it was deposited.

Time-depth graphs

The above dates allow construction of a time/depth model for the 30m core, based on the summary diagram shown in Figure 63. An even spread of dates has been assumed between each dated level, although there will definitely be resulting inaccuracies caused by lack of measured dates, especially in the deep sand levels. There is reasonable likelihood that the organic silt deposits will have accumulated at a relatively steady rate over the duration of deposition, but this is unlikely to apply to the sand horizons. There will inevitably have been periods when large quantities of sand were deposited in a short period of time, especially during adverse weather conditions, and also that significant depths of sand may have been removed at other times as a result of wind and/or water action.

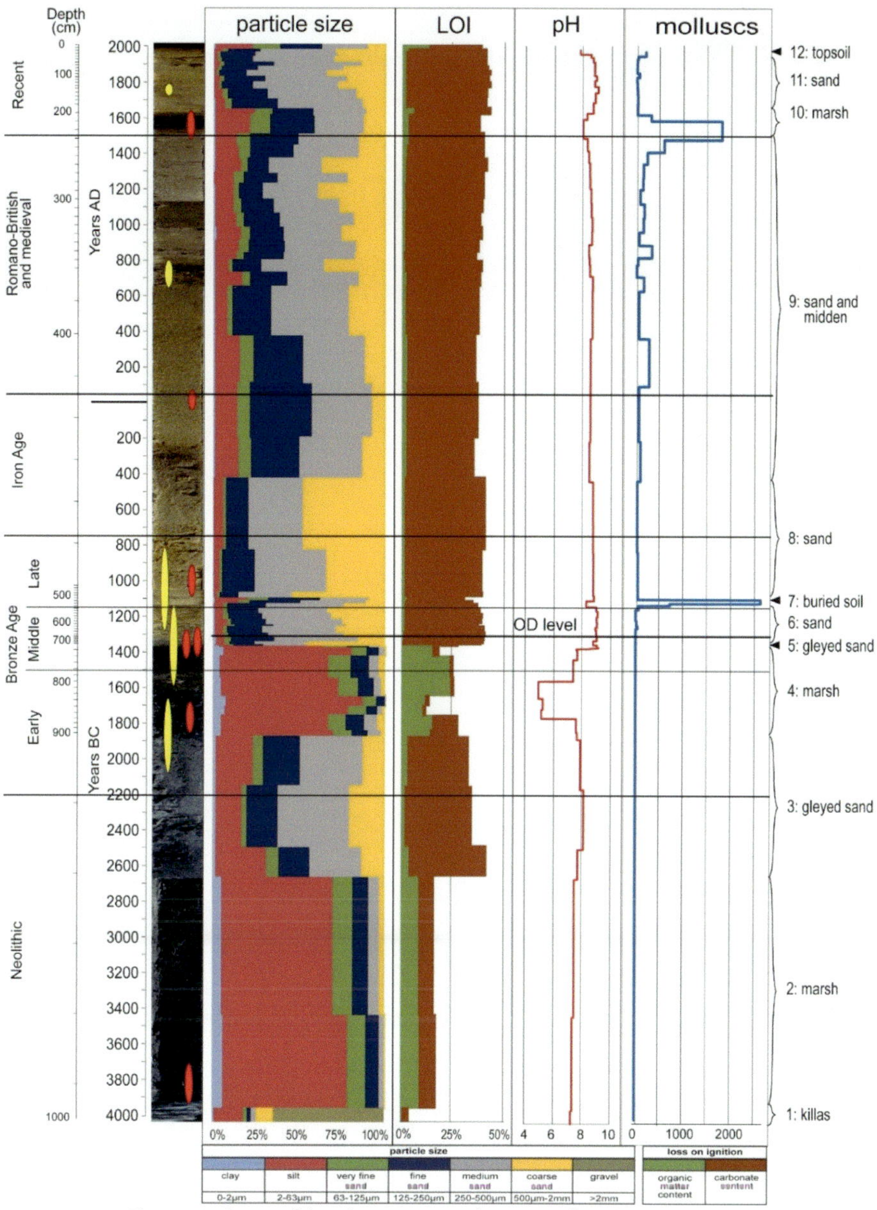

Figure 118. Age/depth summary diagrams for the 30m core.
The dots indicate the dated sediments with 2σ ranges: ○ = OSL, ● = radiocarbon.

Given these limitations, Figure 118 shows the 30m summary diagram modified to show the data arranged on an even chronological basis, as opposed to the even depths on the original diagram A time/depth model can also be constructed for 0.5m, with extrapolation of the dated levels (Figure 119). It is accepted that there will be some error as a result of this extrapolation, but this is likely to be minor, owing the similarity of stratigraphy between the cores.

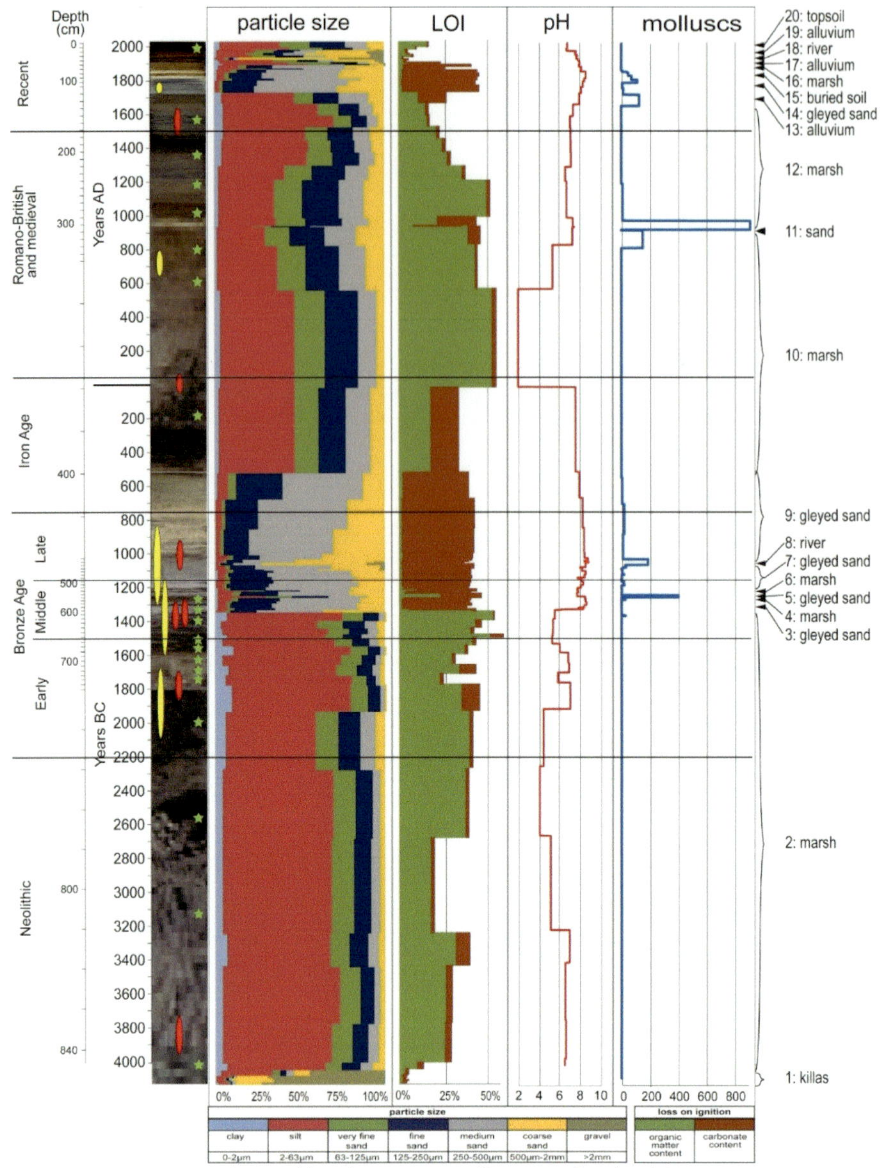

Figure 119. Age/depth summary diagrams for the 0.5m core.
The dots indicate the equivalent levels at which material was dated on the 30m core: ○ = OSL, ● = radiocarbon.
Stars ☆ indicate the level at which pollen samples were obtained.

In the discussion which follows reference is made primarily to the findings in the 30m core, and the deposits numbered in square brackets [ ] are those given in the right column of Figure 118. Evidence is also drawn from the pollen analysis performed on the 0.5m core (Figure 119), which is described in detail in Chapter VIII.

Mesolithic, Neolithic and early Bronze Age

There is abundant evidence that there was human presence during the Mesolithic in that flint scatters have been found over a wide area in this region of Cornwall, but the coring sequences are all post-Mesolithic and give no indication of the palaeoenvironment during the early Holocene. Mesolithic worked flints were found in the trench excavation (Chapter X), so it is clear that there was a Mesolithic presence in the immediate vicinity of the later Bronze Age settlement.

The earliest date, obtained from alder wood in the marsh deposit [2] immediately above the underlying killas, is 3945–3705 cal BC, which represents the commencement of marsh formation at this location. The pollen analysis is consistent with these findings, showing a wetland environment of fen/carr woodland, with alder (*Alnus glutinosa*) and willow (*Salix*) and a range of herbaceous and aquatic taxa consistent with a damp floodplain surface with areas of still, or slowly moving, water. Oak (*Quercus*), hazel (*Corylus*) and birch

(*Betula*) indicate a deciduous woodland/scrubland which was relatively open and perhaps limited in extent. Elm pollen is absent, suggesting either that it did not grow in the area, or that the earliest sediments post-date the early Neolithic elm decline (Parker *et al.* 2002). It is likely that this fen/carr habitat extended across the valley during these early stages of deposition, but the site of the main river channel at this time is not known. Large fragments of alder wood were present in the lowest horizon in all the valley samples, and there was an unburnt hazel nut in the lowest sediment at 15m. The pollen does not include any taxa which might indicate saline conditions, supporting the hypothesis that this was not a marine estuary at the time these sediments were deposited.

Whether this marshy ground extended fully across the valley to the south cannot be stated with certainty as neither of the 'south' cores reached the underlying killas, but it would seem likely. The marsh deposits are currently 220–260cm thick, but may originally have been much thicker, no account being taken of compaction over time as a result of the weight of accumulating deposits. Allen (2000) has demonstrated that this may have a significant influence on the interpretation of stratigraphy in coastal lowlands, with differential autocompaction across valleys and over hillocks and depressions varying by several metres, although no hillocks are likely in the river basin.

The position of the river during this time is not established, but there is no evidence of any clay sediments which might have indicated estuarine deposits; it is concluded that in the survey area from the early Neolithic this was always a freshwater valley, and at no time was it marine, although the pollen analyses raise the possibility of some salt marsh in the vicinity. This suggests that local sea level during this phase of the Holocene may have been around 3m below modern OD level. However, there may have been a substantial sand bar or dune system, and the presence of some sand through the marsh phase makes this possible. Furthermore, caution is therefore needed when modelling sea level curves (Edwards 2006; Brain *et al.* 2012).

Middle and late Bronze Age

Although there was some sand deposition through the marsh phase of the Neolithic and moderate sand deposition between *c.* 2700 and 1800 cal BC (at the 30m location, but not at 0.5m), the main period when sand appears commenced *c.* 1400 cal BC. The transition from marsh to sand on the south side of the valley seems to concur with the change on the north side.

At 30m the sand was replaced by further marsh [4] at about 1800 BC, securely dated by OSL in the sand (2080–1660 BC) below marsh dated by radiocarbon (1885–1685 cal BC). The pollen indicates a return to wetter conditions about 1750 BC, indicated by decline in alder/willow and increase in sedges/bur reed. Oak pollen values decline about this time, suggestive of a reduction in dry woodland. The marsh was acidic at this time and the occurrence of heather pollen (*Calluna vulgaris*) may indicate a coastal heath community with plantain and cinquefoil (*Potentilla* type).

The chronology of the transition to wetter conditions across the valley and from sand to marsh at 30m corresponds closely to the commencement of the first known settlement at Gwithian, around 1850 cal BC, during the early Bronze Age (Phase 1) (Sturgess 2007a). Although definitive evidence is lacking, Dr Batchelor, in his pollen report, considers anthropogenic interference may be the cause of vegetation change due to the timing of the event in the early Bronze Age: the increase in herbaceous taxa, some of which may indicate disturbed ground, e.g. fat hen (*Chenopodium* type) and ribwort plantain (*Plantago lanceolata*), and relatively high values of microcharcoal. However, the date of this change from sand to marsh closely correlates to the 1880 cal BC major riverine flooding event elsewhere in Britain identified by Johnstone *et al.* (2006) and, if that also affected Cornwall, may have been a contributing factor in the return to marsh conditions.

There is an abrupt change in the sedimentology from marsh [4] to sand [5, 6] during the middle Bronze Age, reflected in all the valley cores. The time at which the marsh retreated and was replaced by sand is securely dated, there being marsh at 1420–1265 cal BC (radiocarbon date), but sand at 1550–1130 BC (OSL date). Despite the wide range in the OSL date it is clear that the latest time that the marsh persisted was 1270 BC and that sand was definitely present by 1130 BC. This corresponds to the later part of the middle Bronze Age levels in the settlement (Phase 3), following which the fields on the west side of the site were partly abandoned (Phase 4), shown in the 2005 trench excavations which contained catholic and shade mollusc with few open country shells (Davies 2006; Sturgess 2007a, 29). Unfortunately, no shells have survived in the core samples from this depth, so no comparison can be made regarding the molluscs at the two sites. The pollen, also, has not been preserved sufficiently for any conclusions to be drawn concerning the vegetation.

Phase 5 of the Bronze Age settlement, from *c.* 1300–900 cal BC (Sturgess 2007a, 30), was a period during which the settlement prospered, with new buildings and a flourishing field system. The sediments in the core indicate a period of stability, being maximal at 1100 BC, during which a buried soil [7] accumulated, sufficient for a very rich and diverse molluscan fauna to develop (18 species, Shannon index: 2.17). There is a mixture of taxa indicating that the ground was still wet (principally *Carychium minimum, Vertigo antivertigo* and *Ashfordia granulata*), but with a good spread of species associated with more varied habitats (mainly *Cochlicopa* sp., *Aegopinella nitidula, Cochlicella acuta, Helicella itala*). The dominant shell is *Lauria cylindracea*, a species requiring some shade, but commonly associated with stone walls, suggesting that there may have been walls in this vicinity, perhaps with adjacent ponding to provide habitat suitable for the wet ground species, similar to modern day conditions behind the road wall (see discussion for 51m core earlier in this chapter). The river channel was mobile during this period, being at the 15m location at *c.* 1350 BC and *c.* 1220 BC, at 0.5m at *c.* 1060 BC and then back to 15m again at *c.* 1050–800 BC.

Iron Age, Romano-British and medieval

Sand [8] covers this buried soil, with an OSL date range 28cm above its base of 1290–810 BC. The thickness of the sand makes it difficult to be precise about the timing of accumulation, but it was certainly about this time, and correlates very well with final abandonment of the Bronze Age settlement around 900 cal BC. It has long been proposed (Thomas 1958) that inundation with sand was the precipitating factor for abandonment, and the present study supports this hypothesis. It would seem likely that the large dune on which the post-Roman industrial site was situated was formed during the late Bronze Age or Iron Age, when significant quantities of blown sand began accumulating on the valley sides for the first time. The deep sand units on the slope of the Towans would seem to be of a similar date.

The sediments in the 30m core from 207–435cm [9] have all been identified as midden material, on the basis of the presence of numerous animal bone fragments and of the mollusc assemblages, both marine and non-marine, and are considered to derive from the post-Roman settlement on the adjacent sand dune, occupied from the fifth to eighth centuries AD (Sturgess 2007b). The age/depth model has produced a misleading impression here, and the midden is likely to cover a far shorter period of time, contemporary with the sub-Roman/early medieval occupation, the earliest dated pottery from this area being cal AD 420–600 (1534±29; OxA-14529) (Hamilton *et al.* 2008).

At one time in the early centuries AD part of the valley base became extremely acid, as low as pH 2, perhaps due to areas of stagnant water. These conditions did not persist and when more alkaline conditions returned, around 800 AD, very high numbers of wet ground molluscs (mainly *Galba truncatula, Pisidium* sp., *Carychium minimum*, Succineidae and *Vertigo antivertigo*) are found, although none indicative of flowing water. This marshland did not extend right across the valley, as there is no evidence of any wet ground in either core south of the present river site. This is consistent with the palynology, which, from *c.* 25 cal BC to *c.* AD cal 1550 shows a reduced woodland cover of both wetland and dryland areas, indicated by a reduction in alder and willow, but increase in herbaceous and aquatic flora, including grasses and sedges. Anthropogenic modification of disturbed land, grassland, crops and associated weeds is suggested by the presence of ribwort plantain (*Plantago*), dandelion (Lactuceae), dock/sorrel (*Rumex acetosa/acetosella*), daises (Asteraceae), Brassica family plants (*Sinapis* type) and possible cereal (Poaceae >40μm). Higher values of heather (*Calluna vulgaris*) pollen, *Sphagnum* moss spores and heath (*Erica*) suggests expanding acidic soils and heathland, again perhaps that of a coastal heath community. There is no indication of a shaded environment in the molluscan record, nor even significant dry ground, so presumably there was considerable variation in the environment in different parts of the valley over this long period of time.

Recent

Marsh [10] covers the midden site following its abandonment and was presumably contiguous with wet conditions closer to the river, but it never extended to the south of the valley. It was not until the seventeenth century that sand again accumulated [11]; the ground remained unstable, reflected in the low number of molluscs, until very recent times with stabilisation and formation of the turf and topsoil now present [12]. The mollusc assemblage in the modern turf at 30m is very rich and diverse, but closer to the river molluscs are virtually absent, and no pollen was preserved in the sample taken at 11cm depth on the 0.5m core. The leet, a short distance north of the present river course, was cut sometime between 1839 and 1850 (see Chapter II). Figure 119 suggests that there was flowing water at 0.5m in the early twentieth century, and whether this indicates slight variation in the line of the leet over time, or, more likely, that the date is slightly late due to inaccuracies in the dating of sediment deposition over the last three hundred years, is unclear. The modern vegetation beside the leet is lush and varied and it is curious that neither pollen nor molluscs are found in the topsoil, despite the neutral pH. Contamination from mine pollution could explain both the lack of molluscs and of pollen.

Meadow

Turning now to the north of the valley and the meadow below the scarp of Godrevy Towans it is clear that the west end of the dune on which the post-Roman buildings are situated is sand for several metres below present ground surface, there being no indication of a solid spur of bedrock on which the sand has accumulated. The sands become thinner further up the meadow, being nearly 500cm deep just north of the dune (100m core), reducing to only 25cm near the scarp (245m core). Thin stabilisation layers are found in the sands up to the position of the field system about two-thirds of the way up the meadow, but generally the sands are unstable, with very few molluscs. Above the field system there are no stabilisation layers.

There is a suggestion of woodland or scrub at the 100m location in the earliest deposits above the bedrock, likely to correspond to the early Bronze Age horizons in the nearby field systems. Interestingly Davies (2006) found, in the 2005 trench, shade species in layers above the basal sand corresponding to the middle Bronze Age. The two analyses, about 200m apart, do suggest that there were shaded areas capable of supporting a variety of shade-loving species (e.g. *Discus rotundatus*, *Clausilia bidentata*) but that its extent and location may have varied over time. It is possible that they relate to periods of abandonment when upkeep of the field systems in the area had lapsed. Spencer (1974, 1975) did not find species suggestive of woodland, but speculated that there may have been woodland clearance prior to the establishment of the Bronze Age settlement.

Further details of the findings at 150m will be given in Chapter X, but it appears that agricultural practices were established here during the Bronze Age. The core within the field system (175m) has sediment throughout entirely consistent with the plough soils revealed in the 1960s excavations. Over the meadow as a whole there is good evidence that there has been grazing but this was intermittent and associated with periods of sediment stability represented by buried soils.

The Towans

Little needs to be said about the sand deposition on the scarp of the towans. On the steep slope there is, somewhat surprisingly, moderate stability at least for part of the deposition period, although at times the sand here is likely to have been continuously lost and replaced by wind/water action, and quantities of sand extracted during the seventeenth to nineteenth centuries. Over the gently rising upper slopes there are hollows in which sediments can stabilise and vegetation grow sufficiently to allow molluscs requiring some shade to thrive. The appearance of marram grass and scrub in more recent times also permits good mollusc populations to develop in areas of resulting stability.

On the flat summit of the towans molluscs thrive on the open ground, with catholic and shade species appearing where sufficient shade is provided by rank vegetation, but with no evidence of woodland. The study suggests that the field systems associated with Crane Godrevy may be at least 50m more extensive than previously thought, with the radar showing one probable ditch, and the molluscs suggesting disturbed ground during, or since, medieval times.

Chapter X - Trench excavation

The transect core samples, because of their small size, only permitted a limited analysis of the sediments through which they passed. There is no opportunity to see stratigraphy *in situ*, either in plan or in profile. The excavation of a small trench allowed the stratigraphy at one location to be examined in detail and gave the opportunity for larger samples for palaeoenvironmental analysis to be obtained. The location at 150m along the transect line was selected as the core suggested that buried soils were present at this site, close to the field system excavated in the 1960s (Fowler and Thomas 1962). There was therefore a high likelihood that there had been human activity in the vicinity both during the Bronze Age and in the post-Roman period. Consent to conduct the excavation was obtained from English Heritage (Scheduled Monument CO 771), Natural England (SSSI) and the landowner, Professor Charles Thomas.

EXCAVATION METHODS

The excavation was conducted over five days in July 2012 by a team from the University of Reading. It was considered that a 2 x 2m trench would provide adequate working space to enable the bedrock to be reached at about 1m depth. The east margin of the trench was located on the transect line, with an area of ground to the west reasonably level and free from obvious recent rabbit activity. The exact co-ordinates of the corners of the trench were plotted using differential GPS:

	Grid references		Elevation
	Easting	Northing	
Northwest corner	SW 58883.97	SW 41285.23	8.12m
Northeast corner	SW 58885.91	SW 42185.24	8.04m
Southeast corner	SW 58885.90	SW 42183.27	8.02m
Southwest corner	SW 58883.93	SW 42183.28	7.92m

The upper portion of turf was removed by spade and placed aside. All sediments below this were sieved using sieves with a mesh of 10mm or ½in; sieving with a smaller mesh size was attempted but the sands were too wet to permit this. Spoil was placed on groundsheets in separate piles for each context. The excavation was conducted by context, with recording of artefacts, stones, shells, etc. as small finds. All material retained in the sieves was recorded as bulk finds. Animal footprints found on the buried soil horizons were photographed and traced on plastic sheeting at life size.

The underlying natural geology was reached at an average depth of 100cm. A small (20 x 20cm) pit was dug in the north-west corner of the trench to obtain samples of the natural and to provide working space to insert the sample tube for OSL dating. On completion of the excavation section drawings at 1:10 scale were prepared from all four trench sections.

Samples for palaeoenvironmental analysis were obtained from several locations within the trench (Figure 120). Five samples for OSL dating were taken by Dr Helen Roberts (Aberystwyth University) from the south facing section to date the old ground surface and the sands within, above and below two buried soil levels. Samples for mollusc and other geoarchaeological analyses were taken from a column at the west end of the south facing section and a 50cm monobloc sample covering the buried soils obtained between the mollusc column and the OSL sample positions (the monolith column has not been analysed, but has been retained as an archive). Finally, two small samples for micromorphology were taken from possible ard-marks in the west facing section.

After completion of the recording and sampling the trench was backfilled by hand, with the spoil being returned as far as possible in context sequence, and the turf replaced.

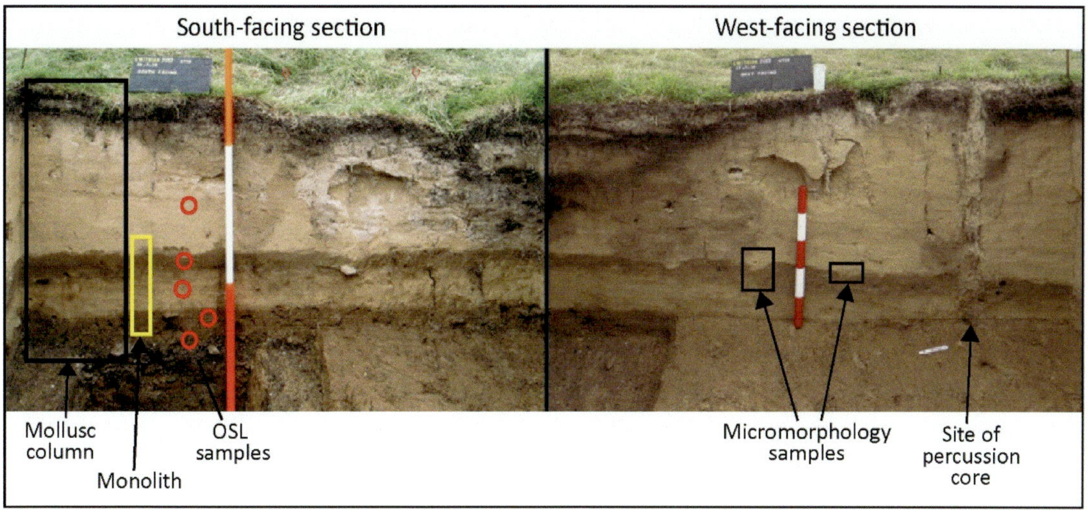

Figure 120. The positions of the dating and environmental samples obtained from the trench.
Scales — left: 50cm divisions; right: 10cm divisions.

EXCAVATION FINDINGS

The drawing of the south- and west-facing sections of the trench is shown in Figure 121; numbers in square brackets [ ] in the following text refer to the contexts shown on the section drawing. Munsell numbers for soil colour of the sediment colours were recorded in the field and are shown in Table 30.

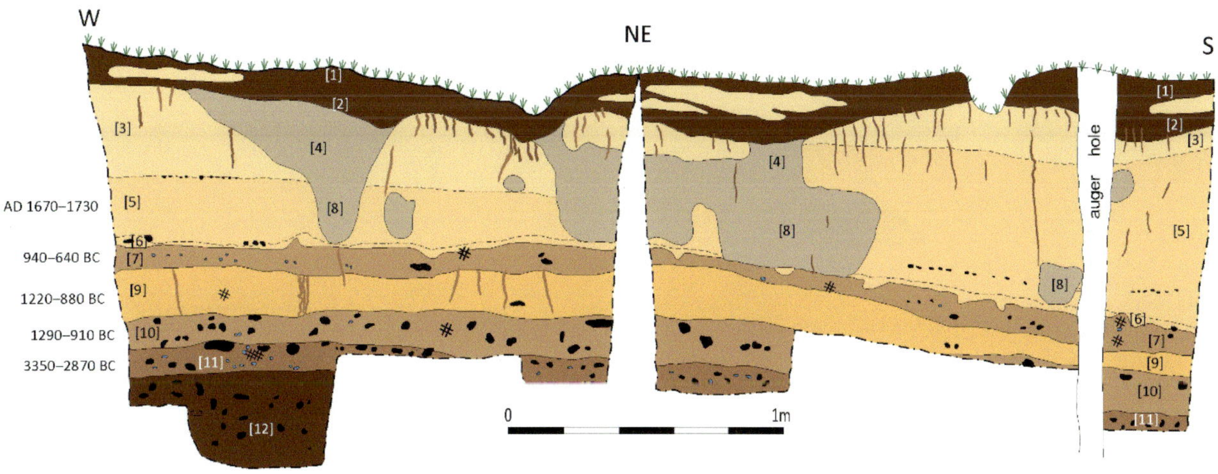

Figure 121. Drawing of the south-facing and west-facing sections of the trench excavation.
The ard-marks cut into the upper surface of [7] are seen, particularly in the west-facing section.
The dates are OSL dates (see later in this chapter).

The turf [1] and sub-soil [2] was 10–20cm thick and the only artefacts within it were some clay pipe fragments. Beneath the turf rabbit burrows became evident [4, 8], but limited to part of the trench only. Within the disturbed sand of the rabbit burrow [4] there were several rabbit bones, some sandrock and one decaying golf ball. None of these were considered of importance; there is no record there has ever been a golf course in the vicinity.

The turf overlay a 35cm thick sand layer [3] above 20cm of slightly darker sand [5] (measured at the north west corner), with a horizon of small stones at the interface. No finds were present in either of these sands apart from a few mussel shell fragments in the lower part. The lowest 4cm [6] was indistinguishable in colour, but clearly different in that it contained numerous fragments of sandrock (1.44kg) up to about 6cm in diameter. Only four fragments of sandrock were found in any other context (apart from the rabbit burrows), two fragments each in contexts [5] and [9] and sandrock is clearly a feature principally of context [6]. *Mytilus* sp. fragments (9g) were also found in this horizon, as well as several small stones up to 6cm diameter, weighing in total 0.5kg. Other marine molluscs within context [6] were two *Nucella lapillus* (dog whelk) shells and one *Patella pellucida* (blue-rayed limpet) shell; this latter is important as it is a small limpet which lives in the holdfasts of Laminarian fronds, and is likely to have been brought to the site with seaweed.

A horizon of sand with silt consistent with a buried soil [7] was clearly visible, being darker in colour; it contained numerous stones (28.1kg), mostly less than 6cm in diameter, but with one fragment 15cm in length. The stones were mainly the local Cornish killas of slate and quartz. The marine molluscs found in this layer were *Mytilus* sp. fragments (344g) and *Patella* sp. (two shells). There were several small charcoal fragments, the largest being 16mm in length. This buried soil horizon will be discussed in more detail below.

There is then another pale sand layer [9] containing stones (16.8kg) of similar size and shape to those in the layer above, again mostly slate with some quartz. The marine shells were *Mytilus* sp. (116g), *Patella* sp. (five shells), *Nucella lapillus* (one fragment) and *Littorina obtusata* (one shell), together with one barnacle plate. Some charcoal fragments were present, but considerably fewer than in the overlying buried soil.

A second buried soil horizon [10], similar to the one above, again contained numerous stones (19.7kg), a few of which were fire-cracked. Only one stone was clearly beach-worn. The stones included eight flint fragments, of which four had evidence of working (Figure 122). The marine molluscs were *Mytilus* sp. (85g), *Patella* sp. (31 apices) and *Pecten* sp. (one fragment). Only very few charcoal fragments were found in this sediment.

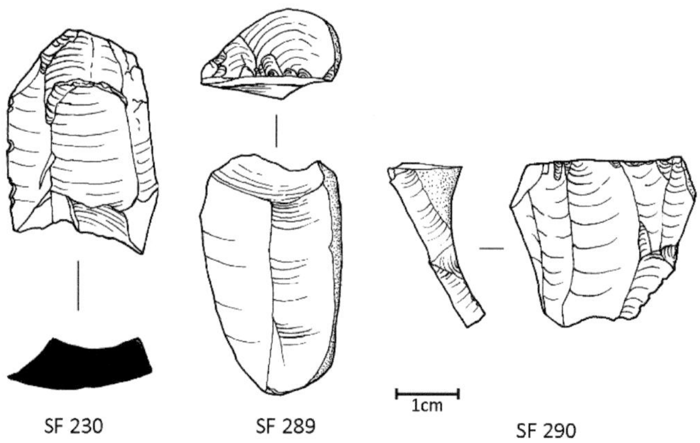

Figure 122. Three of the worked flints from the lower buried soil [10] in the Trench. (graphic: Jennifer Foster)

The old ground surface [11] is a silty sand with several charcoal fragments, some stones (7.1kg) including two worked flint fragments. There were some marine molluscs: *Mytilus* sp. (27g), *Patella* sp. (26 apices) and *Nucella lapillus* (one fragment). The flint fragments from this and the overlying context are considered to be of Mesolithic origin (Charles Thomas, pers. comm.).

The bedrock of killas [12] was only excavated in the north-west corner of the trench to allow working space for the OSL sample collection and mollusc sampling at the base of the trench.

Buried soils

The upper buried soil horizon [7] was dark in colour but with numerous marks of lighter coloured overlying sand impressed into its upper surface (Figure 123), shown in close-up in Figure 124, and diagrammatically in Figure 125. These were mainly semicircular in shape, ranging in size up to 10cm in diameter, and are interpreted as animal hoof prints. The large prints, over 7cm, are likely to be bovid, while those less than 5cm are probably ovi-caprid (Bell 2013, Plate 7.2). All these species have cloven hooves but only a few of the prints in the trench were clearly cloven, the majority being semicircular. However, when footprints are impressed into soft sediment the cleft posteriorly is often lost, as is clearly demonstrated in the prints at Redwick (Bell 2013, 146). Laboratory examination for the presence of spherulites was undertaken to assess whether animal dung was present, but none were found in samples from any of the buried soil sediments.

Figure 123. The upper surface of context [7] showing the animal footprints and ard-marks (*). (scale: 10cm divisions).

Figure 124. Close up of the footprints (scale: 1cm division).

Figure 125. Drawing of the footprints and plough/ard-marks in context [7].

Two linear stripes, running north-south, were also present and are interpreted as ard-marks. In the cleaned west facing section there were several 'U-shaped' indentations in the upper surface of the buried soil (Figure 126), and micromorphology (Figure 127) supported the interpretation that these were ard/plough marks, although no east-west marks were recognised during the excavation. Chapter XI includes the full micromorphology report by Dr Rowena Banerjea. No spherulites were seen in the micromorphology sections. The slides confirmed the presence of a thin buried soil overlying blown sand. However, other features such as the colour and mollusc assemblages suggest that this entire context is likely to be a buried soil, albeit not so well developed throughout, implying that true stability was perhaps only achieved in its upper portion.

Figure 126. The west-facing section showing indentations consistent with ard-marks on the upper buried soil. (scale: 10cm divisions).

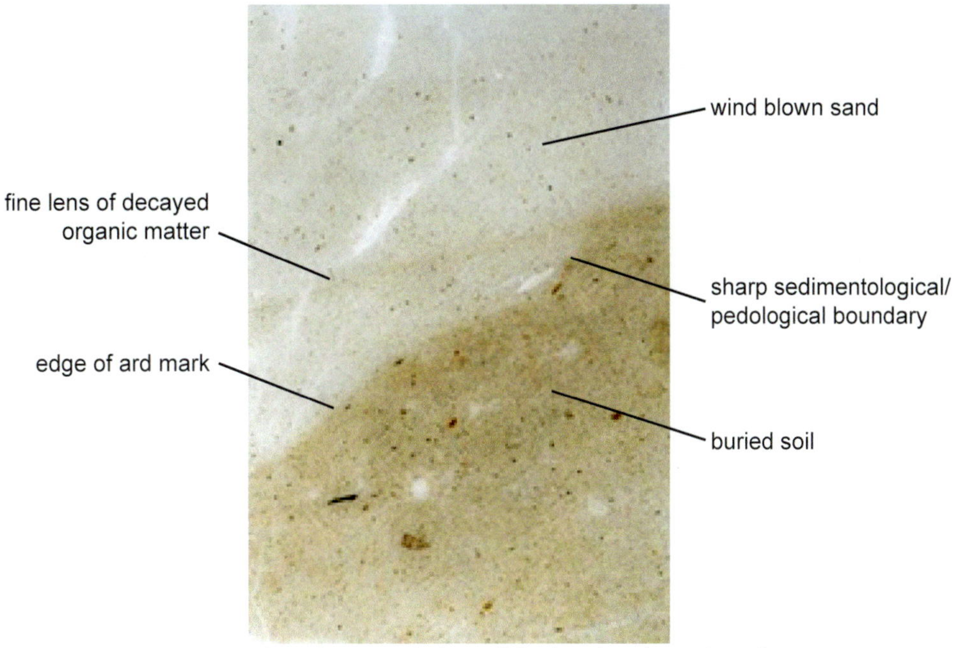

Figure 127. Micromorphology slide showing the ard-mark.
(photo: Dr R. Banerjea)

A very thin (<500μm) lens of decayed organic matter overlay the ard-mark which was considered may represent a possible brief phase of stabilisation over the plough marks, but which did not develop into a soil. Of relevance is that one of the micromorphology slides demonstrated a thin calcrete/sandrock crust on the surface of the buried soil, entirely consistent with the finding of the numerous sandrock clasts in the overlying sand [6].

The lower buried soil [10], in contrast to the upper soil, showed only a very few footprints (Figure 128). These were 10cm in diameter and are consistent with bovid prints. No ard-marks were detected in this horizon.

Figure 128. The hoof prints in the lower buried soil [10].
The ard-marks cut into the upper buried soil [7] are visible at the top of the image (scale: 10cm divisions).

DATING OF THE TRENCH SAMPLES

Five samples for OSL dating were obtained during the excavation (Figure 120) to determine the chronology of the buried soil horizons. In addition, four molluscs from the buried soils, all *Xerocrassa geyeri*, were submitted for radiocarbon dating, one of which (that at 67-70cm depth) was duplicated by the radiocarbon laboratory. The aim of using both dating techniques was to obtain comparison between the methods, and to establish the times at which this particular mollusc was living at Gwithian. The results are shown in Table 28 (OSL) and Table 29 (radiocarbon) and graphically in Figure 130.

Table 28. OSL dates on the mollusc column sand samples.

Lab number	Depth (cm)	OSL date (years before 2010)	Calendar date (1σ)	Mid-date	Context
Aber-203/GWT-12-1	45 ± 3	310 ± 30	AD 1670–1730	AD 1700	[5] overlying sand
Aber-203/GWT-12-2	64 ± 3	2800 ± 150	940–640 BC	790 BC	[7] upper buried soil
Aber-203/GWT-12-3	77 ± 3	3060 ± 170	1220–880 BC	1050 BC	[9] intervening sand
Aber-203/GWT-12-4	87 ± 3	3110 ± 190	1290–910 BC	1100 BC	[10] lower buried soil
Aber-203/GWT-12-5	97 ± 3	5120 ± 240	3350–2870 BC	3110 BC	[11] old land surface

Table 29. Radiocarbon dates on the mollusc column *Xerocrassa geyeri* shells.

Lab number	Depth (cm)	$\delta^{13}C$	Radiocarbon date (BP)	Calibrated date (2σ)	Mid-date	Context
OxA-28965	60–64	-6.46‰	3599 ± 28	2030–1890 cal BC	1960 BC	[7] top of upper buried soil
OxA-28964	67–70	-6.23‰	3537 ± 28	1950–1770 cal BC	1860 BC	[7] bottom of upper buried soil
OxA-28963	67–70	-6.53‰	3515 ± 29	1925–1750 cal BC	1828 BC	
OxA-28962	85–88	-6.38‰	3576 ± 29	2025–1785 cal BC	1905 BC	[10] top of lower buried soil
OxA-28961	91–94	-7.22‰	3558 ± 29	2015–1775 cal BC	1895 BC	[10] bottom of lower buried soil

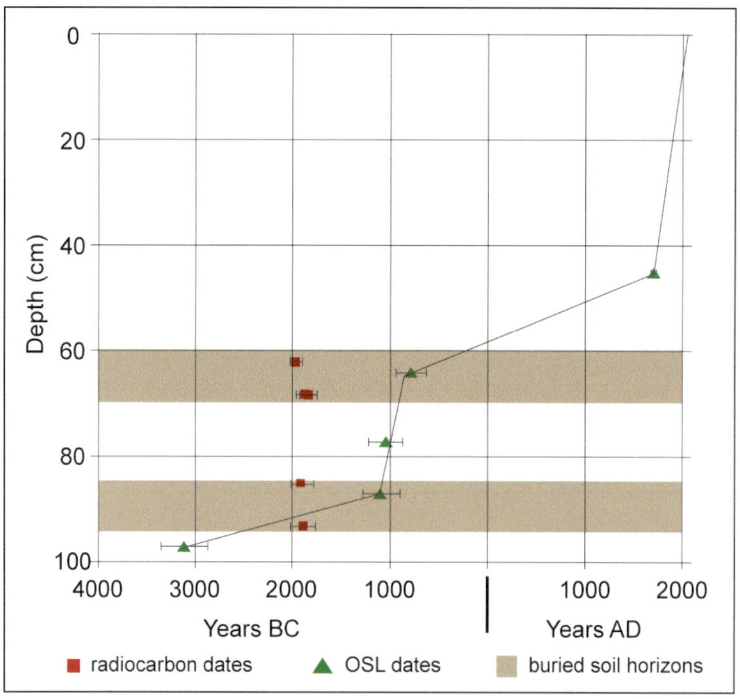

Figure 129. Age/depth plot for the OSL and radiocarbon dates in the trench. The trend lines join only the OSL dates.

It is evident from the age/depth plot that there is discrepancy between the OSL and radiocarbon dates. The OSL dates give a good chronological sequence for the old ground surface, the two buried soil horizons with intervening sand and the overlying sand, and are likely to be a more accurate representation of the chronology than the radiocarbon dates, the latter giving almost identical dates for both of the buried soils. The $\delta^{13}C$ values for the molluscs, except for the lowest one, are very similar which implies that the shells in these three upper samples may actually derive from the same deposit. It is likely that the shells have been moved, most probably as a result of deflation leading to reworking of shells, although animal disturbance cannot be excluded (earthworm burrows were evident in the section, but upward movement of the shells by worms would not be expected).

MOLLUSC COLUMN

The column was taken from the north-west corner of the trench in an area that was clear of rabbit disturbance (Figure 120). Twenty five samples were obtained, one of which was the killas underlying the old ground surface. The laboratory analyses are summarised in Figure 130. The dates shown on the left are the OSL dates, the radiocarbon dates not being shown for reasons explained above.

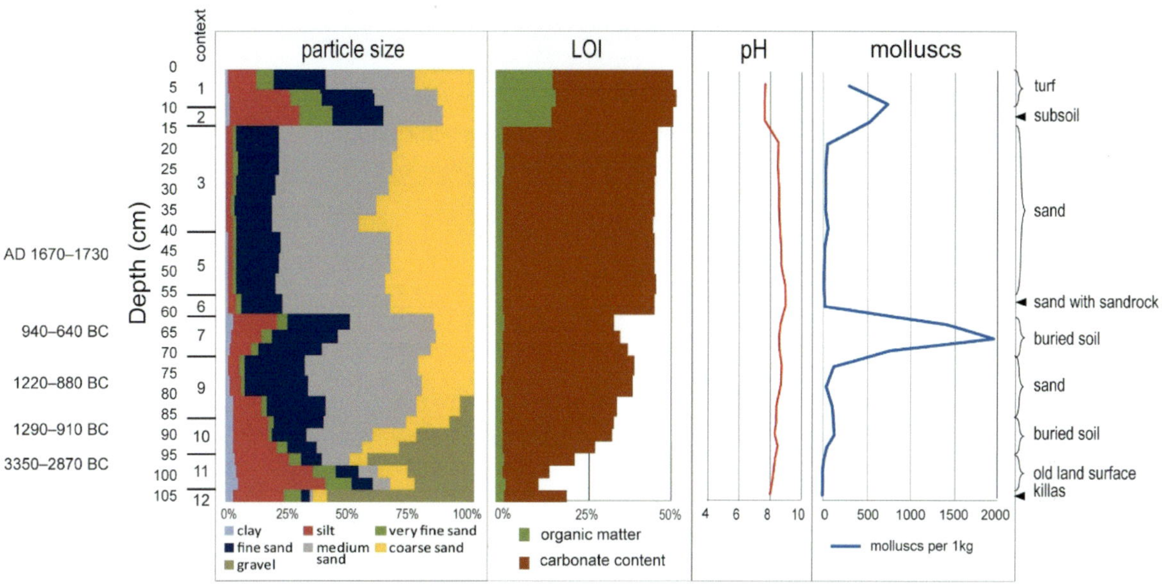

Figure 130. Summary diagrams for analyses on the trench mollusc column.

The killas bedrock [12] contains a moderate proportion of clay/silt (24%) and carbonate (18%), but no molluscs were found in this sediment. The old ground surface [11] did contain a few molluscs, although the carbonate content decreased to 9%, while there was increase in clay/silt to 40%. The proportions of silt and clay steadily decrease during accumulation in the lower buried soil [10], while sand and carbonate content increase, as do the numbers of shells present, although still relatively few in number. The sandy layer [9] between the two buried soils is identified in the particle size analysis and contains only a small number of molluscs, consistent with rapid accumulation.

There is marked change in the upper buried soil [7], with a greater proportion of silt and a very dramatic rise in the numbers of shells present. The progressive increase in silt in this soil indicates a steady build-up of finer grain size over time which is a mark of the increasing stability of this horizon, in accord with the micromorphology findings. The silt in the buried soils is likely to be derived from colluviation from fields upslope, where negative lynchets are cutting into subsoil. The sand above the buried soil [6, 5, 3] has almost no silt and very few shells. There is no clear distinction on these graphs of context [6], no sandrock nodules being present in the sample from the mollusc column. There is some differentiation between the sand layers [5] and [3], with a higher proportion of coarse sand marking the base of [3], the proportion decreasing upwards. The sub- and topsoil [2, 1] is clearly identified both on the particle size and loss on ignition graphs, being more silty and containing a higher proportion of organic matter than the underlying sand, and with a moderate number of molluscs. Horizon [2], is likely to be the result of colluviation.

THE MOLLUSCS

The mollusc findings are shown numerically in Table 30, and graphically in Figure 131. A small number of species indicating open ground dominate the sequence, with a larger range of taxa, some more catholic in their ecological preferences, found in the upper buried soil [7] and the topsoil [1 and 2].

No shells were found in the sample of killas [12]. Very few were present in the old ground surface [11] probably due to relative decalcification; the taxa that do remain indicate open conditions with no evidence of shade in the vicinity. Some small flint flakes were present as well as charcoal. Marine shells in this deposit were *Mytilus* sp., *Patella* sp. and *Nucella lapillus*. The OSL date of this layer is mid-Neolithic (3350–2870 BC).

The findings are sufficient to indicate human presence at this time, perhaps with middening or manuring to account for the marine shells.

The lower buried soil [10] contains a mollusc assemblage of low diversity, with a single species, *Xerocrassa geyeri*, accounting for 92% of all shells. This is a species associated with open, dry, calcareous environments and its abundance in the soil to the exclusion of almost all other taxa indicates that it was one of the first molluscs to establish itself in thin blown sand accumulating on the old ground surface during the middle Bronze Age. The few other molluscs show that there were some areas of a mixed environment providing some shade. *Mytilus* fragments are present in moderate numbers, as well as a few *Patella* fragments.

Table 30. The molluscs from the Trench column.
1kg samples. No shells were found in the basal killas (>103cm).

	0-5	5-9	9-14	14-20	20-26	26-31	31-36	36-40	40-45	45-50	50-55	55-60	60-64	64-67	67-70	70-75	75-80	80-85	85-88	88-91	91-94	94-97	97-100	100-103
Context	1		2	3					5			6	7			9			10			11		
Sediment	turf		sub-soil	sand					sand				buried soil			sand			buried soil			old land surface		
Munsell number	7.5YR 3/3		10YR 5/2	10YR 7/6				10YR 6/4	10YR 7/4				10YR 5/4	10YR 4/4		10YR 6/4			10YR 5/4			10YR 4/3		
Carychium minimum																1		1		1				
Cochlicopa lubrica													40	26	25	7		3	2	3		1		
Cochlicopa lubricella			1										3	3										
Cochlicopa spp.		3						1					31	42	10	1			1	3	2			
Vertigo pygmaea	41	63	18										28		1							1		
Pupilla muscorum	2	2	22	1	6	3	5		1	4		2	309	230	148	28	6	4	1	3				
Lauria cylindracea																					1			
Vallonia excentrica	163	378	180	6	2	4		1	1	1	1	1	316	331	228	38	8	7	2	1		1	1	1
Punctum pygmaeum	23	28	8			1							14	2										
Vitrina pellucida	2		19			1	4										1			2				
Vitrea contracta													1											
Aegopinella nitidula													9								1			
Limacidae	1	1	6																					
Clausilia bidentata													1											
Cochlicella acuta	34	214	224	26	15	12	12	9	11	6	7	10	632	1228	230	25	2			1		2		2
Ashfordia granulata	1		15										2											1
Candidula intersecta	4	8	14	1	2																			
Cernuella virgata	9	19	9	10	6	1	2	27		2	4	1												
Helicella itala	3	4	2	9	4	13	10	18	12	7	5	10												
Xerocrassa geyeri							1	3				2	43	123	138	27	26	99	120	121	49	14	2	1
Trochulus hispidus	16	31	17	1																				
Cepaea nemoralis													1			1								
Cepaea sp.																			1					
Cornu aspersum	1	1						3																
TOTAL SHELLS	**300**	**752**	**535**	**54**	**35**	**35**	**34**	**62**	**25**	**20**	**17**	**26**	**1430**	**1985**	**780**	**128**	**43**	**114**	**127**	**135**	**53**	**19**	**3**	**5**
TOTAL SPECIES	13	12	13	7	6	7	6	7	4	5	4	6	13	7	6	7	5	5	4	7	4	5	2	4
Shannon index	1.54	1.39	1.62	1.43	1.54	1.50	1.53	1.43	0.97	1.43	1.23	1.38	1.51	1.17	1.52	1.61	1.12	0.55	0.30	0.52	0.35	0.93	0.64	1.33
Brillouin index	1.47	1.36	1.57	1.27	1.33	1.28	1.53	1.28	0.82	1.17	1.00	1.14	1.49	1.16	1.50	1.52	0.99	0.50	0.25	0.46	0.28	0.71	0.37	0.82
Mytilus sp. (umbos)									frag			1	frags	1	frags	frags	1	1	6	4	9	11	6	4
Patella sp.										frag				frag					frags		frags	2	frags	frag
Nucella lapillus																frag							frags	1
Charcoal														(+)	+		(+)	+	++	+++	++	+++	+++	++
Flint flakes																					1		1	3

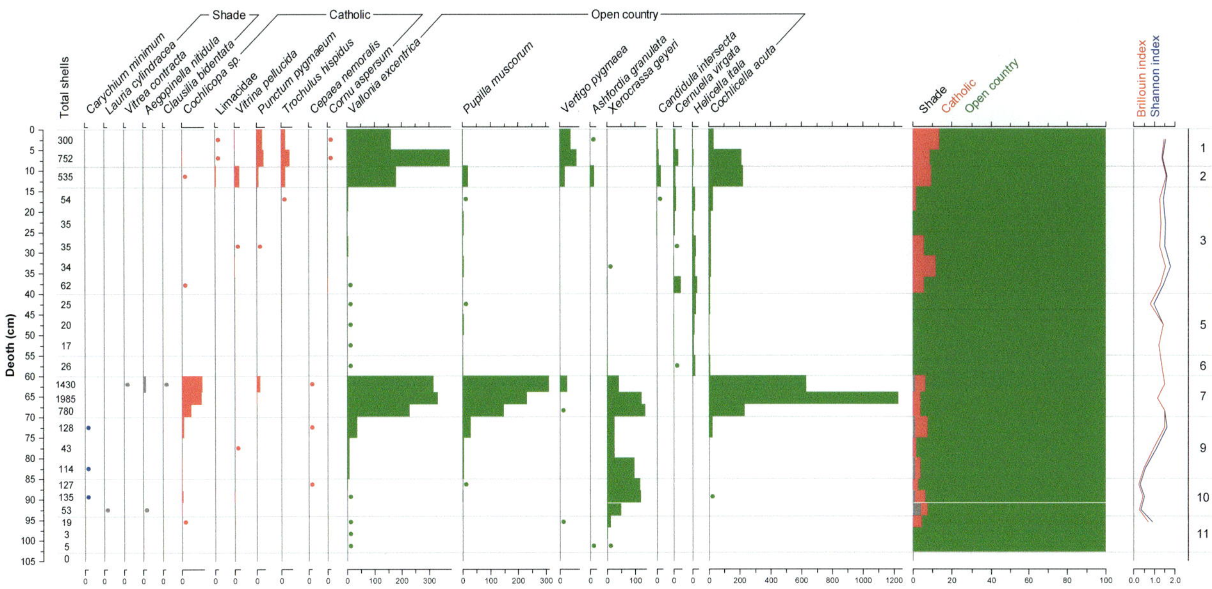

Figure 131. Mollusc diagram for the Trench samples.
(absolute numbers of molluscs; dots represent single specimens)

The molluscs in sand horizon [9], between the buried soils, is also dominated by *Xerocrassa geyeri*, with few other shells, although *Pupilla muscorum*, *Vallonia excentrica* and *Cochlicella acuta* become more prominent in the upper third of this layer. These are likely to represent the early stages of stabilisation, but may all be derived from the overlying buried soil, moved downwards by worm or rain/wind action.

The upper layer of buried soil [7], dated to the late Bronze Age/early Iron Age, is very rich in shells, with 780, 1985 and 1430 per kg from bottom to top, and has high mollusc diversity. *Cochlicella acuta* is now the dominant species, with corresponding decline in *Xerocrassa geyeri*. *Pupilla muscorum* and *Vallonia excentrica* are strongly present, the former being an early coloniser of newly cleared ground, the latter at these levels of abundance being indicative of dry open grassland (Evans 1972, 266), probably grazed (Cameron and Morgan-Huws 1975; Evans and Evans 1995). *Vertigo pygmaea*, another species of short grassland, appears in the upper part of this sand. A moderate number of more catholic molluscs, mainly *Cochlicopa lubrica*, although with some *Punctum pygmaeum*, are also found in these soils. *Aegopinella nitidula* is able to live in the shade found at the base of long grass, as shown in the modern turf blocks at several locations (see Chapter V). The single *Clausilia bidentata* is an anomaly, and is probably intrusive from woodland/scrub in the vicinity, brought here by animal or wind action.

The sands overlying this buried soil [6, 5] contain very few molluscs indicating an unfavourable habitat with rapid accumulation and/or regular deflation. *Helicella itala* is first encountered; it is perhaps surprising that this species is not found in lower levels, as it is considered to be a native species and not a late Holocene introduction. *Cernuella virgata* first appears, albeit in small numbers; if current opinion suggesting this is a probable post-Roman introduction is correct, then its presence gives support to much sand in this horizon having been lost to deflation. The upper part of the thick layer of sand [3] does contain a few more shells, mostly open country. *Candidula intersecta*, considered a likely medieval introduction, is found only in the uppermost portion of this sand.

Mollusc numbers increase markedly as the more stable subsoil and modern turf is reached. The mainly open country assemblage is entirely consistent with the modern habitat of grazed grass and tussocks of taller vegetation. Several catholic species are present (Limacidae, *Vitrina pellucida*, *Punctum pygmaeum*, *Trochulus hispidus*, *Cornu aspersum*), but no shade species and it is probable that grazing by the abundant rabbit population and intermittent cattle and sheep has kept much of the meadow vegetation sufficiently cropped so that shade species cannot easily survive. The decline in *Vallonia* and *Cochlicella* in the most superficial layer may, however, be associated with the gradual increase in tussocky grass with patches of scrub that is currently developing, as revealed by comparison with photographs of the 1960s excavations (Nowakowski *et al.* 2007, Colour plate 1) with those of the present landscape.

DISCUSSION

The finding of flints dated to the Mesolithic is entirely consistent with flint scatters from this period previously found over the wider area of the Gwithian/Godrevy landscape (Thomas 1958; Roberts 1987; Sturgess 2004; Thomas 2007), although none have previously been reported from the area of this trench.

The OSL date for the old ground surface (3350–2870 BC) suggests that sand first accumulated on the killas hillside during the middle Neolithic. Neolithic Gabbroic pottery sherds were found in the original Bronze Age settlement excavations (Quinnell 2007a), but none were discovered during the present excavation. The strong dominance of the mollusc *Xerocrassa geyeri* in this horizon implies that this was bare open ground during the mid-Neolithic; if the area had been wooded during the Mesolithic then this had all been cleared prior to sand deposition.

The OSL dates for the trench horizons suggest that the lower buried soil formed during the middle to late Bronze Age (1290–910 BC), while the upper buried soil is dated to the late Bronze Age/early Iron Age (940–640 BC). Comparison can be made with the dating of the trench opened in 2005 within the confines of the Bronze Age settlement (Hamilton *et al.* 2008), and from which 20 radiocarbon and two OSL dates were obtained. Figure 132 shows a section of the 2005 trench with the OSL dates marked, to compare with the equivalent section in the new trench, shown in Figure 133. Both sections show two dark buried soil horizons separated by a layer of paler sand. It is easy to assume that these are equivalent coeval deposits in the two sections. There is, however, discrepancy in the OSL dates. It is noted that the dates for the 2005 trench (Table 31) are chronologically inverted, although there is overlap in date range and they were accepted as valid (Hamilton *et al.* 2008).

Figure 132. Section of the 2005 trench with OSL dates. (modified from Roberts 2006, Figure 12)

Figure 133. Section of the 2012 trench with OSL dates. (photo: Martin Bell)

Table 31. Correlation of the chronology of the 2005 and 2015 trenches. The phases are those established during the original Bronze Age site excavations (Sturgess and Lawson Jones 2006a).

2005 trench		Period	Phase	2012 trench		Dates
		Medieval		Overlying sand	AD 1670–1730	OSL
		Late BA/early IA	5	Upper buried soil	940–640 BC	OSL
		Late BA	5	Intervening sand	1220–880 BC	OSL
		Middle/late BA	5	Lower buried soil	1290–910 BC	OSL
Overlying sand	1805–1485 BC		4			
Upper buried soil		**Middle BA**	3			
Intervening sand	1515–1195 BC		2			
Lower buried soil		**Early BA**	1	Molluscs	2030–1750 cal BC	^{14}C
		Neolithic		Old ground surface	3350–2870 BC	OSL

The lower buried soil horizon in the 2005 trench was considered to equate to the Phase 1 early Bronze Age level of the original excavations (*c.* 1800 cal BC) and this is consistent with the new dates obtained on molluscs in the old ground surface. The upper soil in the 2005 trench was considered to equate to original Phase 3 middle Bronze Age level (*c.* 1500–1200 cal BC) (Nowakowski *et al.* 2006), for which there is no equivalent in the new trench.

The soils in the 2012 trench showed two phases, the lower dated to the earlier part of the late Bronze Age (1290–910 BC), and the upper to the later part of the late Bronze Age (940–640 BC), with an intervening episode of sand blow (1220–880 BC). All of these horizons fit chronologically with Phase 5 of the original excavations which were dated to *c.* 1300–900 cal BC (Sturgess 2007a, 30). These periods correlate well with the evidence of agriculture found previously, although it was suggested that arable cultivation was less intensive during this phase (Sturgess 2007a, 32). It is now proposed that the focus of agriculture moved west to the area of the field system excavated in the early 1960s (Fowler and Thomas 1962), although that excavation did not explore the prehistoric horizons.

The late Bronze Age ard-marks and evidence of animal hoof prints indicate that this area of land was both in arable and pasture use. Ard-marks found in the Bronze Age layers during the original excavations (Megaw *et al.* 1960–61; Thomas 1970; Sturgess 2007a, 30) were identified both at the base of buried soils as impressions cut into the underlying sand (dark lines cut into yellow sand) and also on the upper surface of soils (lighter lines cut into darker buried soil (shown in Figure 19). Ard-marks were only recognised in the present study as being cut into the upper surface of the buried soil.

Prehistoric ard-marks (Fowler and Evans 1967) are not infrequently found in appropriate locations (e.g. Benbecula, Outer Hebrides: Shepherd and Tuckwell 1976–77; Alveston, Avon: Everton and Fowler 1978; Sumburgh, Shetland: Lamb and Rees 1981; Jura, Inner Hebrides: Stevenson 1984; Brownslade, Pembrokeshire: Groom *et al.* 2011), but the presence of animal footprints in British sites is rare (St Marys, Isles of Scilly: Evans 1984; Formby, Lancashire: Huddart *et al.* 1999; Redwick and Peterstone, Gwent: Bell 2013). These new findings indicate that animal husbandry was practiced certainly during the later Bronze Age occupation phase at Gwithian (on the upper buried soil), when footprints are abundant, and perhaps during the middle Bronze Age (on the lower buried soil). Animal movement may have been transient, as there was no evidence of spherulites to indicate the presence of dung.

The mollusc assemblages can assist in understanding land use during the phases in which the buried soils accumulated. The lower soil, likely middle to late Bronze Age, contained very few shells other than the open country species *Xerocrassa geyeri*, which it has been suggested derive from an earlier, deflated, horizon. This xerophilic species is found today on dry, open, calcareous ground with short vegetation, rocks or rock rubble, in natural or undisturbed habitats. The virtual absence of *Vallonia excentrica* implies that short grazed turf was not a feature at this time. This is in marked contrast to the upper soil, which is likely late Bronze/early Iron Age. Although there are still substantial numbers of *X. geyeri* (which may be reworked from lower levels), *Cochlicella acuta* is now the dominant species, and there are large numbers of *Pupilla muscorum* and *V. excentrica*. The ground is now much more stable, and the strong presence of *Vallonia* (22% of all shells) shows that this was short grazed grassland, entirely consistent with the finding of animal hoof prints in this horizon.

There is no evidence of any soil horizon or agricultural activity relating to the post-Roman phase of occupation of the site, despite this being found in the excavations within the nearby ridge-and-furrow field system only 20m uphill from the 2012 trench (Fowler and Thomas 1962). No scientific dating was undertaken during those 1960s excavations, but the lower layers produced pottery sherds typical of the early and middle Bronze Age as well as numerous sherds from more superficial layers with a typology suggesting ninth to late tenth century AD. No pottery of any type was found during the 2012 excavation, with the conclusion that the post-Roman activity did not extend this far down the meadow, or that the port-Roman horizons/fields have been lost to deflation.

Fowler (1962) postulates that manuring took place on the post-Roman field systems, indicated by the inclusion of shells, animal bones and teeth, pebbles and sherds within the plough-soil. He considered that tiny mussel and periwinkle shells, too small for foodstuff, and various little beach pebbles, pointed to the use of seaweed as manure, in addition to midden material. Although beach pebbles were not a feature of the present excavation (apart from the single stone in context [10]), the large quantities of mussel shells, together with the *Patella pellucida* shell in the layer just over the upper buried soil, point to seaweed being used as manure on the site of this trench during the late Bronze Age. Although Bell (1981) cautions against assuming that such materials derive from seaweed specifically brought in as manure, as they may have come from other sources such as domestic middens or farmyard manure, the presence of *P. pellucida*, a small shell which lives in seaweed holdfasts and with no cultural use, is good evidence of seaweed manuring at Gwithian. The use of this resource as manure has been well attested at other sites in Cornwall (Carew 1766, 27) and elsewhere (Smith 1994; Bakels 1997; Guttmann *et al.* 2005; Smith 2012, 390), and the present finding adds to the corpus of knowledge.

Sometime during the late Bronze Age or early Iron Age blown sand covered the buried soil, and the very small number of shells found in the overlying sand implies continual instability. The thin sand horizon immediately covering the soil contains large quantities of sandrock and stone and it is probable that one or more episodes of winnowing removed higher layers of sand leaving heavier grains. Where the sandrock originated is uncertain, as this is found mainly in the cliffs about 1km to the northwest, although there is one surface outcrop in the meadow about 200m west of the trench. Sandrock has for long been used as building material in this area of Cornwall (Bristow 2011; King 2011) and it may be that there were sandrock field walls which became decayed due to stormy weather or neglect.

Sand continued to accumulate and remained unstable until the modern sub- and topsoil became established. An OSL date at 45cm depth (about one-third up these sands) is AD 1670–1730, meaning that there is only 15cm of sand representing over 2000 years of deposition, and it must be supposed that there have been regular episodes of winnowing over this period, probably leaving bare dune for much of the time. Only in much more recent times has stability returned to the area, with establishment of good vegetation cover and recolonisation by molluscs, reflected by the high numbers of both shells and of taxa, albeit with some reduction in the uppermost layer, perhaps related to the gradual return of the meadow to taller vegetation and scrub.

Chapter XI – Micromorphology analysis of a buried soil.

by Dr R. Y. Banerjea, Quaternary Scientific, University of Reading

INTRODUCTION

The profile in section (Figure 134) showed two buried soil horizons, separated by 15cm of blown sand, with the upper soil layer showing numerous animal footprints on its upper surface and probably plough marks (Figure 123). Micromorphology samples were collected from the west-facing section of the excavation trench, from the upper soil layer to analyse the plough marks in more detail, to identify any evidence for cultivation practices (Macphail 1998; Lewis 2012) and to understand the soil forming processes and post-depositional alterations at the site.

METHODS

Three micromorphology slides were analysed (115m x 75mm) from two monoliths collected from the section face (Figure 134). Two micromorphology slides were prepared from sample 1, (GW1a, b), and one from sample 2 (GW2).

Figure 134. The location and details of the micromorphology samples from the west-facing section of the trench. (scale: 50cm).

SAMPLE PREPARATION

The procedure followed the University of Reading standard protocol for thin section preparation. Samples were dried to remove all moisture and then impregnated with epoxy resin while under vacuum. The impregnated samples were left overnight so that the resin entered all of the pores. The samples then are placed in an oven to cure for 18 hours at 70°C before they are ground and polished on a Logitech LP30 precision lapping machine to the standard geological thickness of 30μm.

SAMPLE DESCRIPTION

Micromorphological investigation was carried out by using a Leica DMEP polarising microscope at magnifications of x40 – x400 under Plane Polarised Light (PPL), Crossed Polarised Light (XPL), and Oblique Incident Light (OIL). Thin-section description was conducted using the identification and quantification criteria set out by Bullock *et al* (1985) and Stoops (2003), with reference to Courty *et al* (1989) for the related distribution and microstructure, Mackenzie & Adams (1994) and Mackenzie & Guilford (1980) for rock and mineral identification, and Fitzpatrick (1993) for further identification of clay coatings. Tables of results use the descriptions, inclusions and interpretations format used by Matthews (2000) and Simpson (1998). Photomicrographs were taken using a Leica camera attached to the Leica DMEP microscope.

RESULTS AND INTERPRETATION

Micromorphology descriptions for each deposit are recorded in Table 32, the frequency and types of inclusions within these deposits are recorded in Table 33, and the abundance of post-depositional alterations and pedofeatures within the deposits is recorded in Table 34. To determine the deposit type classification, each deposit was grouped using the following diagnostic sedimentary attributes and inclusions which provide crucial information concerning the origin of inclusions, transportation mechanisms of particles and the deposition processes. To ascertain the origin of sediment components descriptions were made of particle size, shape, and the composition of the coarse and fine fraction, particularly the frequency of rock, minerals and anthropogenic inclusions (Table 33). The depositional events are characterised by the following sedimentary attributes: sorting, related distribution, orientation and distribution of the inclusions (Table 32), and bedding structure (Table 33). Understanding the formation processes for deposits is crucial to interpreting the depositional pathways of rock fragments and minerals, any anthropogenic debris such as charred wood and artefacts, and other types of plant remains and microfossils (Schiffer 1996; La Motta and Schiffer 1999; Matthews 2010). Analysis of post-depositional features provides crucial information concerning the effects of weathering, preservation conditions (Bisdom *et al.* 1982; Courty *et al.* 1989; Canti 1999; Brady and Weill 2002; Breuning-Madsen *et al.* 2003) and stratigraphic integrity of the deposit (Courty *et al.* 1989; Canti 2003).

Table 32. Deposit type descriptions to characterise depositional events and materials.

Building/ Feature	Deposit type	Sample number	Unit	Basal boundary	Particle size	Sorting	Fine material	Groundmass	Colour	Related distribution	Microstructure	Inclusions orientations and distribution
Probable cultivation horizon	Wind blown sand	GW1a	1	Sharp, smooth pedological/ sedimentological boundary	Sand: medium sand (200-500µm). Occasional coarse sand (500µm-1mm) components	Well sorted	None	Crystallitic – high birefringence	PPL: grey with brownish orange flecks. XPL: white/grey with high order interference colours	Monic	Simple packing voids 40%. Channels 10%	Locally orientated basic and clustered
	Buried soil	GW1a	2	N/A	Sand: medium sand (200-500µm). Occasional coarse sand (500µm-1mm) components	Moderately sorted	Organic/iron stained	Crystallitic – high birefringence	PPL: grey with brownish orange flecks: XPL: white/grey with high order interference colours and dark orange flecks	Linked and coated	Complex packing voids 20%. Channels 5%	Moderately-strongly oriented, basic, mostly horizontal and parallel
	Buried soil	GW1b	2	Smooth, gradual sedimentological boundary	Sand: medium sand (200-500µm). Occasional coarse sand (500µm-1mm) components	Moderately sorted	Organic/iron stained	Crystallitic – high birefringence	PPL: grey with brownish orange flecks. XPL: white/grey with high order interference colours and dark orange flecks	Linked and coated	Complex packing voids 20%. Channels 5%	Moderately-strongly oriented, basic, mostly horizontal and parallel
	Wind blown sand	GW1b	3	N/A	Sand: medium sand (200-500µm). Rare coarse sand (500µm-1mm) components	Well sorted	None	Crystallitic – high birefringence	PPL: grey with brownish orange flecks. XPL: white/grey with high order interference colours	Monic	Simple packing voids 40%. Channels 10%	Moderately-strongly oriented, basic, mostly horizontal and parallel, but getting more locally orientated, basic and clustered towards the base
	Wind blown sand	GW2	1	Sharp, smooth sedimentological boundary	Sand: medium sand (200-500µm). Occasional coarse sand (500µm-1mm) components	Well sorted	None	Crystallitic – high birefringence	PPL: grey with brownish orange flecks. XPL: white/grey with high order interference colours	Monic	Simple packing voids 40%. Channels 5%	Locally orientated basic and clustered
	Calcrete crust	GW2	4	Sharp, smooth sedimentological boundary	Sandy clay: medium sand sand (200-500µm) in a clay matrix	Well sorted	Mineral	Crystallitic	PPL: greyish brown/orangey brown. XPL: pale yellowish brown/orangey brown	Embedded and coated	Channels 5%. Spongy 5%	Moderately-strongly oriented, basic, mostly horizontal and parallel, but occasional locally oriented, basic and clustered areas
	Wind blown sand	GW2	5	Sharp, smooth sedimentological boundary	Sand: medium sand (200-500µm). Occasional coarse sand (500µm-1mm) components	Well sorted	None	Crystallitic – high birefringence	PPL: grey with brownish orange flecks. XPL: white/grey with high order interference colours	Monic	Simple packing voids 40%. Channels 10%	Mostly locally-moderately oriented, basic, in clustered
	Buried soil	GW2	2	Sharp, smooth sedimentological boundary	Sand: medium sand of different compositions	Moderately sorted	Organic/iron stained	Crystallitic – high birefringence	PPL: grey with brownish orange flecks: XPL: white/grey with high order interference colours and dark orange flecks	Linked and coated	Complex packing voids 20%. Channels 5%	Moderately-strongly oriented, basic, mostly horizontal and parallel
	Wind blown sand	GW2	3	N/A	Sand: medium sand of different compositions	Well sorted	None	Crystallitic – high birefringence	PPL: grey with brownish orange flecks. XPL: white/grey with high order interference colours	Monic	Simple packing voids 40%	Moderately-strongly oriented, basic, mostly horizontal and parallel, but getting more locally orientated, basic and clustered towards the base

Table 33. The frequency and types of inclusions present.

Deposit type	Sample number	Unit	Thickness on slide (cm)	Bedding	Rock fragments and mineral inclusions										Bio-archaeology	Plant remains			
					Flint	Limestone	Mn/Fe replaced	Quartz	Feldspar: microcline	Mica: Muscovite	Mica: Biotite	Manganese	Glauconite	Earthworm granules	Shell	Charred plant remains	Plant tissue (with cellulose)	Plant tissue (no cellulose)	Plant tissue: ?calcified
Wind blown sand	GW1a	1	3.0-7.5	Massive		*	*	**							******			*	
Buried soil	GW1a	2	2.5-7.5	Massive	**			**			*			*	*****	**			*
Buried soil	GW1b	2	1.5	Massive	**			**			*			*	*****	**			*
Wind blown sand	GW1b	3	7.5	Massive	*		*	**			*				******			*	
Wind blown sand	GW2	1	2.4-3.2	Massive	*			**	*	*	*			*	******				
Calcrete crust	GW2	4	0.351.5	Massive	**			***							****		*		
Wind blown sand	GW2	5	0.1-1.0	Massive	*			**		*	*			*	******				
Buried soil	GW2	2	2.5	Massive	**		*	**		*	*	*	*	**	*****			*	
Wind blown sand	GW2	3	0.5	Massive	*		*	**		*	*				******			*	

Key: ****** = >70%’ ***** = 50-70%; **** = 30-50%; *** = 15-30%; ** = 5-10%; * = <5%

Table 34. The abundance of post-depositional alterations and pedofeatures.

Deposit type	Sample number	Unit	Weathering				Decay	Bioturbation	
			Translocation		Chemical alteration			Microstructure effects	Excremental pedofeatures
			Dusty impure clay coatings: unlaminated	Iron	Manganaese neomineral formation	Secondary calcite	Organic staining	Mesofaunal/root bioturbation	Earthwork granule
Wind blown sand	GW1a	1					*	***	
Buried soil	GW1a	2	**	*			**	**	**
Buried soil	GW1b	2	**	*			**	**	**
Wind blown sand	GW1b	3					*	***	
Wind blown sand	GW2	1		*				**	**
Calcrete crust	GW2	4		**		*****	*	**	
Wind blown sand	GW2	5						***	**
Buried soil	GW2	2	**	*	**		**	***	***
Wind blown sand	GW2	3							

Key: ***** = >20%; **** = 10-20%; *** = 5-10%; ** = 2.5%; * = <2%

Characterisation and formation processes

Five depositional units (Figure 135 & Figure 136) have been identified within the samples. Units 1, 3 and 5 are classified as windblown sands, unit 2 as a buried soil and unit 4 as a 'calcrete crust formation'. Units 4 and 5 only occur in slide GW2.

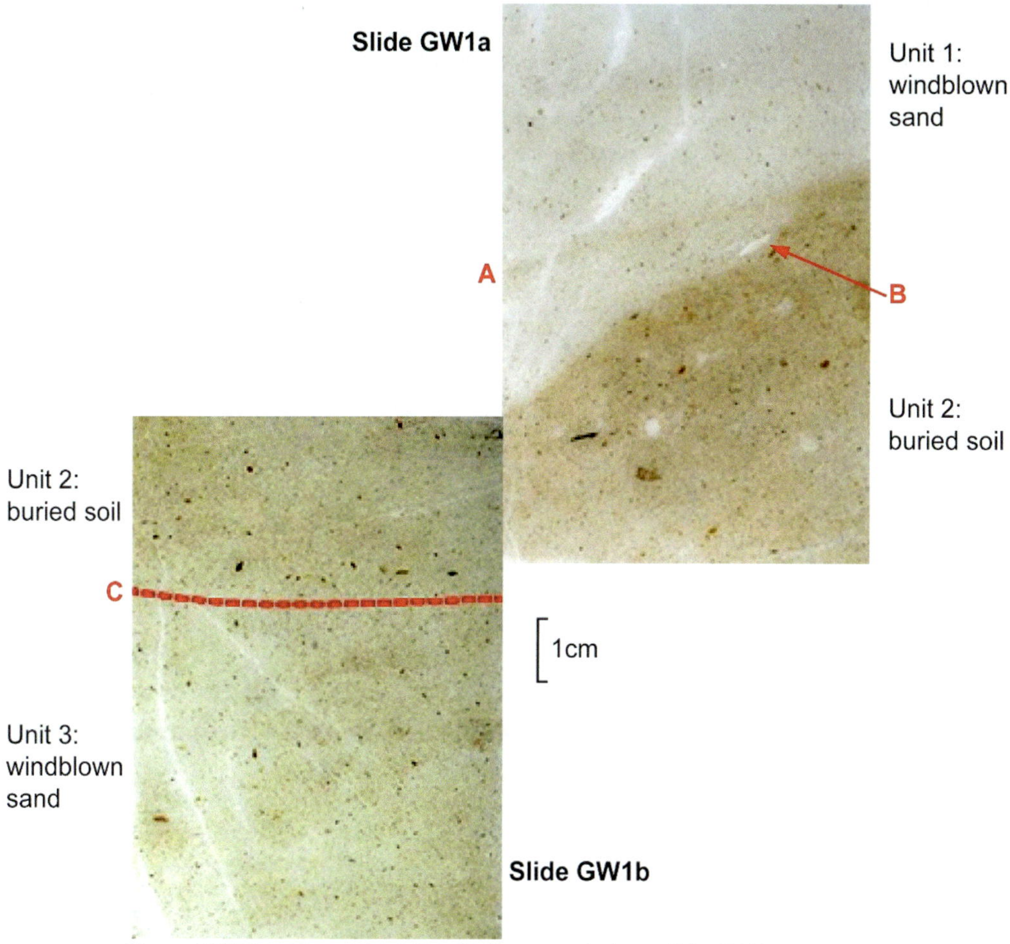

Figure 135. Stratigraphic units of micromorphology slide GW1.

A: fine lens of decayed organic matter;
B: smooth, sharp sedimentological/pedological boundary;
C: smooth, gradual sedimentological boundary.

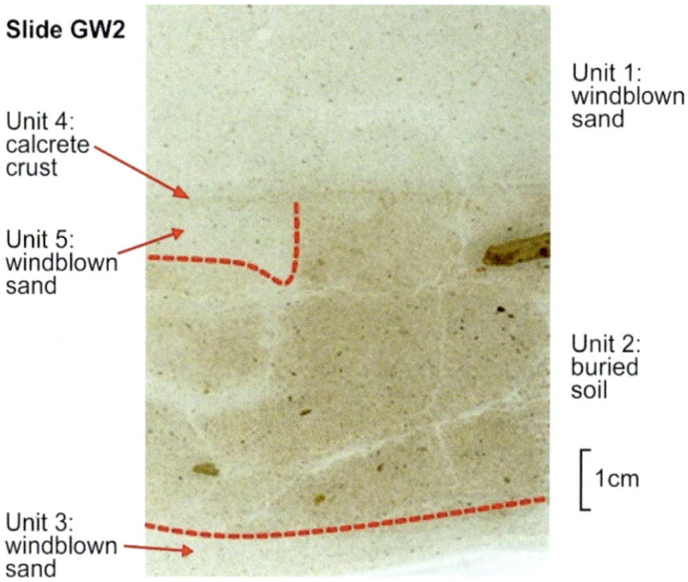

Figure 136. Stratigraphic units of micromorphology slide GW2.

Windblown sands (Units 1, 3 and 5)

Units 1, 3 and 5 share typical classification characteristics with coastal sand dunes in south-west England and Wales (Bell and Brown 2008, 18): medium grained sands (200–500µm), with occasional/rare coarse grains (500–1000µm), well sorted with grains of different compositions (Table 32). These units have a monic related distribution, an absence of fine material and simple packing voids (40%) in the microstructure. In origin both units predominantly comprise calcareous inclusions, 70%, of shell, fossils and limestone fragments (Table 32) including tiny complete shells and other marine organisms (Figure 137) (Bell and Brown 2008, 19). Other inclusions consist of angular/sub-angular flints (<5%), quartz grains (<15%), feldspars (<5%), mica (muscovite and biotite, <5%) (Table 33). Inclusions within dunes are not always the product of weathered bedrock geology but may include material recycled from, for example, earlier dunes, cover sands and outwash gravels (Bell and Brown 2008, 18). The quartz and feldspar inclusions may be the product of weathered granitic areas of Devon and Cornwall (Bell and Brown 2008, 19).

Figure 137. Shell and microfossil fragments in GW1a windblown sand (unit 1) (XPL).

Sand dunes form in relatively low energy environments where there is a supply of sand grade sediment (Bell and Brown 2008, 18). Micromorphological characteristics of low energy environments are preferred orientation of inclusions, indicating slow processes and moderate to well sorted deposits. Sediments formed by mass movements of material, such as colluvium, are typically more poorly sorted (Goldberg and Macphail 2006), and on settlements, those deposited by human activity as a result of maintenance are generally unsorted and unorientated and unrelated inclusions show haphazard deposition (Matthews 1995; Goldberg and Macphail 2006; Banerjea 2011). In units 1, 3, and 5 inclusions are locally oriented and related to each other in clusters (Table 32), as if several grains accumulated together; a process repeated throughout the deposit.

Within unit 1 in slide GW1a a thin lens, <500µm in thickness, of decayed organic matter (Figure 138), <5%, occurs. This lens may represent the beginnings of a stabilisation horizon in the dune system that did not take hold.

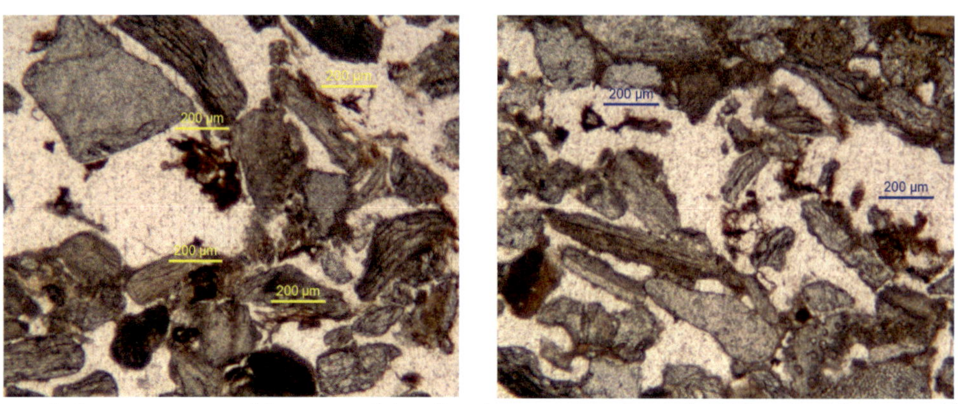

Figure 138. Organic fragments comprising a very faint, fine, lens running through GW1a windblown sand (unit 1) (PPL).

The basal boundary of unit 1, between units 1 and 2 (the buried soil), in slide GW1a and between units 5 and 2 in slide GW2 is sharp, smooth pedological/sedimentological boundary on a 45º angle (Figure 135–B). There is a very clear sedimentary distinction between units 1 and 2.

Buried Soil (Unit 2)

The buried soil is represented in slides GW1a, GW1b and GW2. Identifying palaeosols in the field alone is often extremely difficult because horizons are either missing because of erosion or homogenised by later processes and therefore now invisible. Therefore, a suite of characteristics is used to identify buried soils which include: the nature of burial; characteristic features in the field and in thin section which suggest soil formation or pedogenesis is taking place such as structure, horizon development and clay coating; and the use of biological indicators such as phytoliths, pollen, beetles and molluscs which can give an idea as to development in past environments and in different periods (French 2003).

The buried soil is thin and very poorly developed. The thickness of unit 2 ranges from 1.5cm in slide GW2 to 8.5cm, in slides GW1a and b, taking into account the overlap created during preparation (Table 33). However, there are micromorphological features indicating that pedogenesis took place. Unlike the windblown sand units, the buried soil contains organic and iron-stained fine material which links and coats the coarser sand components (Table 32). The buried soil has 20% complex packing voids which indicates porosity, but also shows that there is a reduction in void spaces in comparison with the windblown sand units. Earthworm granules (Figure 139) occur at a greater frequency than in other units (10% in unit 2, 5% in units 1 and 5, slide GW2, and 5% in unit 2, 0% in units 1 and 3 slide GW1a, b; Table 32) suggesting that there was a greater degree of biological activity in the buried soil than in the windblown sands. It is possible that earthworm granules in the windblown sand deposits (units 1 and 5, slide GW2) have been reworked by more recent mesofaunal activity, evident by channels in the microstructure (Table 32).

Figure 139. Earthworm granule within buried soil (XPL), GW1a buried soil (unit 2).

In comparison with the windblown sand deposits, the sorting and orientation/distribution of inclusions of the buried soil is different. The buried soil is moderately sorted; there is an increased frequency of coarser sand components compared with windblown sands which are well-sorted. The inclusions in the buried soil are more strongly oriented and aligned with the basal boundary, in comparison with locally oriented windblown sands (Table 32), as if these particles have accumulated with lower energy during a period of stabilisation, or were pushed down and aligned horizontally by compaction from ploughing.

The basal boundary of the buried soil (the boundary between units 2 and 3) is a smooth gradual sedimentological boundary (Table 32), suggesting that the soil developed on unit 3 (Figure 135–C). In addition, the buried soil shares a similar rock and mineral assemblage with the underlying windblown sand deposits (Table 33) which strengthens this argument.

Unlike the windblown sands the buried soil contains charred plant remains (Figure 140 and Table 33) and an increased frequency of angular/sub-angular shaped, sand sized (500µm) flint fragments.

Figure 140. Charred plant remains (PPL), GW1a buried soil (unit 2).

Calcrete Crust Formation (Unit 4)

The processes by which calcium carbonate precipitates and accumulates in soils are complex and analysis of their structure using submicroscopic techniques (e.g. scanning electron microscopy) is required to understand their formation processes to a greater extent (Durand *et al.* 2010). However, thin-section analysis shows that, in slide GW2, dense, fine, carbonate material has infilled the packing voids between coarser medium sand rock and mineral components (Table 33) by coating the inclusions to form a cement lens, unit 4, that is 0.35–2.5mm in thickness (Figure 141).These formation processes have resulted in an embedded and coated related distribution, where the coarser rock and mineral components appear embedded within the matrix and coated by the calcium carbonate fine material (Table 32). Carbonate coatings appear to have two main origins: (1) unrelated to biological activity, formed by evaporation or mechanical translocation, the latter with admixes of non-carbonate material and (2) related to biogenic activity, for example coatings consisting of microbial tubules or certain needle-fibre calcite types (Durand *et al.* 2010, 154). Non-carbonate material, specifically iron and organic staining (Table 34), is present causing the fine material to have brownish orange colour in addition to the white/grey calcite which could suggest that carbonate coatings have formed as a result of mechanical translocation.

Figure 141. Calcrete crust (XPL) in unit 4, GW2; note the embedded and coated related distribution of this thin unit.

Interestingly, the calcrete crust, unit 4, occurs at approximately the same depth as the lens of organic matter in unit 1, slide GW1a (Figure 135 and 136).

Post depositional alterations

All deposits show disturbance by weathering, decay processes and mesofaunal and root bioturbation at differing frequencies (Table 34).

Weathering and decay processes

Overall, unit 2, the buried soil, showed more evidence for weathering and decay processes than the other units, specifically dusty impure clay coatings, iron translocation, manganese neomineral formation, organic staining, bioturbation and a higher frequency of earthworm granules (Table 34). Darkening in colour, known as organic staining, is observed in thin-section from the decomposition of organic matter (Courty *et al.* 1989) and shows the decay of organic matter, probably from plant tissue which was identified in thin-section, some of which has become calcified (Table 33).

Unlaminated, dusty impure clay coatings, 2–5%, only occur in unit 2 (Table 34). If the clay coatings were microlaminated, this would suggest that a slow aqueous deposition of fine material had occurred (Courty *et al.* 1989). Dusty impure clay coatings are associated with more turbulent hydraulic conditions (Courty *et al.* 1989). The dark colour of dusty clay coatings is mainly caused by finely dispersed organic matter, and these coatings mainly correspond to the youngest phase of clay illuviation (Kühn *et al.* 2010). Iron staining occurs in most units but most abundantly, 2–5%, in the calcrete crust, unit 4 (Table 34). Free iron is highly mobile only when present in the ferrous state which occurs under anaerobic conditions (Courty *et al.* 1989). Iron has coated void spaces and minerals, channels and minerals.

Manganese neomineral formation occurs in the buried soil, slide GW2 (Table 3). Fluctuating alterations of reducing and oxidising conditions can lead to manganese formation (Brammer 1971; Brown 1997; French 2003). Most critical redox happenings occur in areas where the O_2 supply is partially restricted either by limited aeration or a predominating electron supply. Most of these regions are redox interfaces such as meeting points between roots or microbial surfaces and the soil surface, aggregates and soil pores, sediments and free water, the boundary between organic and a mineral horizon (Bartlett 1988), and, with iron, within horizons at the base of cultivation (Lewis 2012, 91).

Bioturbation

All deposits show evidence of bioturbation in the microstructure in the form of channels and chambers (Table 34). Bioturbation most probably results from either root or mesofaunal activity. It is possible that the earthworm granules that occur within the windblown sand deposits have been reworked by bioturbation from the buried soil into the windblown sand deposits, particularly as, in the windblown sands, earthworm granules, characteristic of those from *Lumbricus terrestris* (Canti 2007) occur close to the boundary with the buried soil.

DISCUSSION AND CONCLUSIONS

Evidence for cultivation

Ploughing

Figure 134-A shows that the morphology of the marks cut into the soil (U/V-shaped, <10cm wide, 1.5–3cm deep) the distance of spaces between them (15–20cm) is consistent with those attributed to ard ploughing through experimental archaeology (Lewis 2012, 36). In comparison to windblown sand units, the buried soil also shows increased silty clay, iron and manganese accumulation (Table 34), accumulation processes which have also been noted within archaeological and experimental cultivation horizons at the base of tilling and at the edges of ard-marks (Lewis 2012, 55, 91-93). However, caution must be applied when attributing these pedofeatures to tillage as the upper part of the soil may have been eroded away, and therefore unavailable for comparison.

The buried soil, unit 2, shows a 50% reduction in packing void spaces in comparison with the overlying windblown sands (Table 32), indicating compaction. While this compaction could have occurred as a result of ploughing (Lewis 2012, 32), it should also be noted that compaction is a feature of palaeosols. As a result of burial, open void spaces can be crushed out of existence by compaction (Retallack 2001, 13). Soils buried by archaeological features such as earthworks can also be affected by compaction (Macphail and Crowther 1996; French 2003; Bell 2009).Therefore, there is a possibility that compaction could have resulted from burial by the overlying windblown sands. There is no evidence for mixing as a result of ploughing infilling the plough marks (see Figure 10 in Lewis 2012, 54); however, given the thinness (<7.5cm) of this soil horizon, it is possible that the upper part of the soil, including any infilling of plough marks, has been eroded away.

Use of fertiliser

It is not uncommon for exogenous materials, such as brick, bone, charred wood and ash to be added as amending materials to increase soil fertility (Adderley *et al.* 2010). Arable soils in Cornwall have traditionally been improved by the addition of shell sand, seaweed and animal manures, and micromorphology samples were collected by Dr Erika Guttmann in the 2005/6 excavations at Gwithian to establish whether such practices took place (Nowakowski 2007b, 16); however, no analysis of these has yet been published. The evidence in unit 2 shows charred plant materials within the buried soil (Figure 140) which may suggest the addition of ashes to improve the soil. Dung and calcareous faecal spherulites have not been identified suggesting that manure was not used as fertiliser in this part of the field. Faecal spherulites would be expected to survive, given the calcareous nature of these sediments, unless the pH dropped below pH 7.7 causing spherulites to dissolve (Canti 1999).

Summary of formation processes

The dune system stabilised and a weakly developed soil, unit 2, formed on the windblown sands, unit 3, at the base of the profile (in micromorphology samples 1 and 2). The soil, unit 2, was probably fertilised with hearth ashes and ploughed in order to be used for cultivation. The boundary between the soil, unit 2, and the overlying windblown sands, units 1 and 5, is sharp and smooth. In addition, there is no evidence for mixing as a result of ploughing infilling the plough marks, suggesting that the upper part of the soil may have been eroded away, which is an occurrence in palaeosols (French 2003), and the remaining lower part of the soil and plough marks covered and in-filled with a later deposit of windblown sand, units 1 and 5. It is possible that unit 2 represents the compacted sediment at the cut base, and has survived erosion due to its compaction.

A lens of decayed organic matter and a calcrete crust, unit 4, formed at the same depth. It is possible that this lens and calcrete crust represent a point in which the dune system began to stabilise again, but that this process did not take hold to develop into a soil.

Chapter XII – Mollusc studies in the wider Gwithian landscape

Several molluscan studies have subsequently been conducted in the wider landscape, and allow the palaeoenvironment of the surrounding area to be explored (Figure 142). Two were in the vicinity of the main archaeological sites but not connected to them, one at the extreme western end of Godrevy Towans, adjacent to the road leading along the top of the cliff, and two on the cliff face adjacent to Strap Rocks, on the southern side of Godrevy Bay.

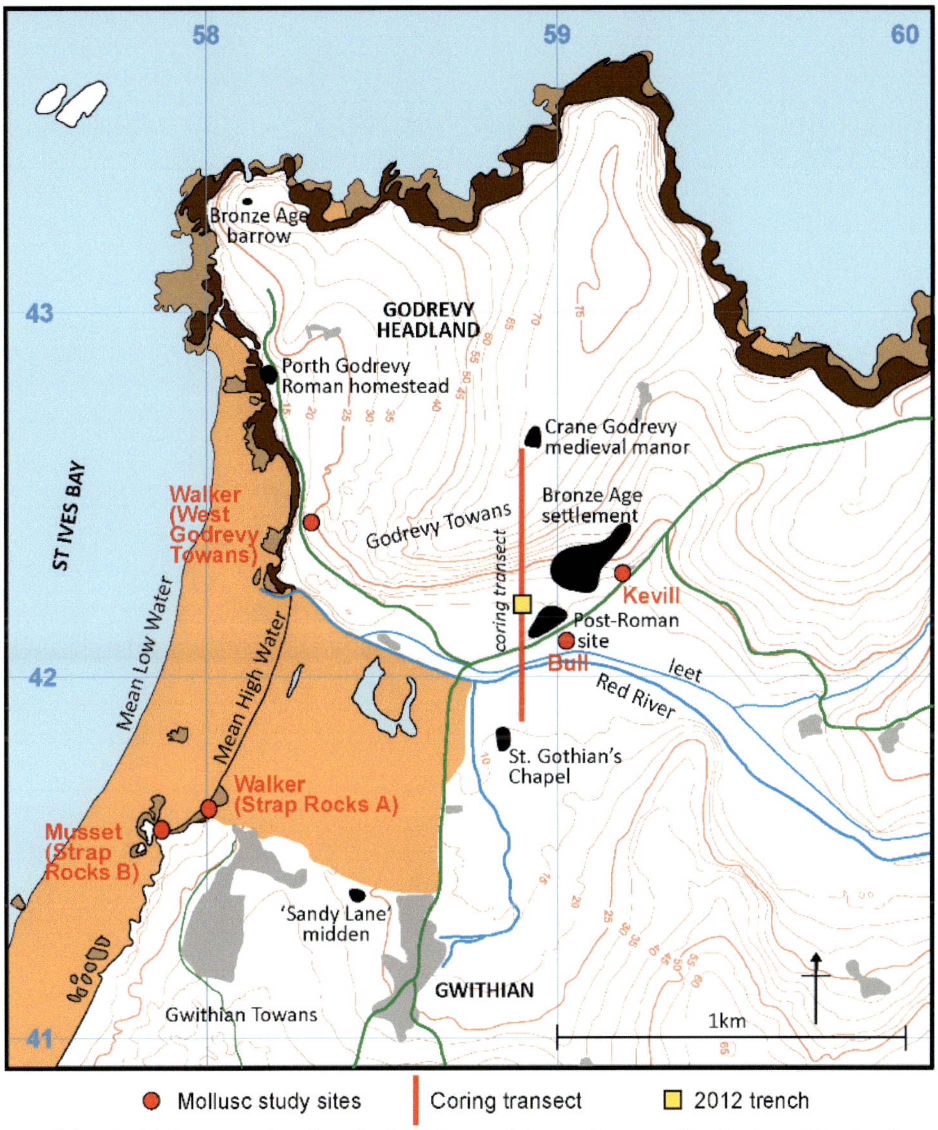

Figure 142. Map of the Gwithian area showing the locations of the mollusc studies in the wider landscape around the main archaeological complex.

West Godrevy Towans – Walker

A mollusc column was obtained at the extreme west end of Godrevy Towans (SW 58286.42457) was the subject of a dissertation by the writer (Walker 2010); a pilot study had revealed abundant molluscs in stabilisation horizons in an exposed scarp and a 3.5m column was investigated in detail (Figure 143). The old ground surface was exposed 1m below the base of the scarp, and there were six stabilisation layers above this, each interspersed with blown sand horizons. The mollusc diagram is shown in Figure 144. Although only a single radiocarbon date was obtained at that time (from charcoal at the base of the old ground surface) further radiocarbon dates on shells have subsequently been obtained (Table 35). The analysis of the charcoal was performed in 2010 by Beta in Miami and that of the shells in 2014 at Oxford, and the exact agreement of the dates for the charcoal and the deepest shell strongly supports the reliability of the shell dating and that the latter has not been influenced by old carbon effect.

Figure 143. The exposure at West Godrevy Towans showing six stabilisation horizons above the old surface (*).

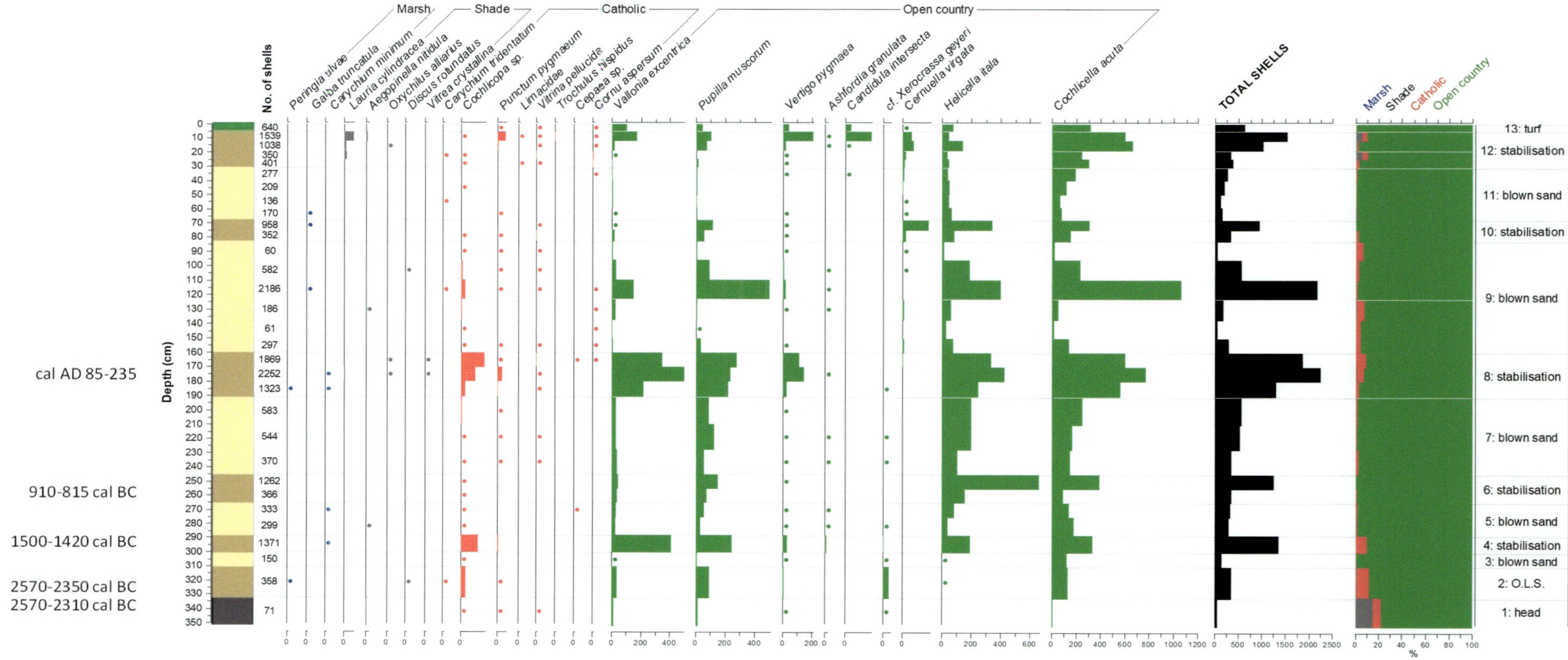

Figure 144. Mollusc diagram for the mollusc column from Godrevy Towans.
Absolute numbers of shells (dots represent ≤ 5 specimens; summary table shows %).
Stabilisation horizons are shown in brown on the stratigraphic diagram, intervening blown sand in yellow.

Table 35. Radiocarbon dates on the Godrevy Towans mollusc column samples.

Lab number	Material	Depth (cm)	$\delta^{13}C$	Radiocarbon date (BP)	Calibrated date (2σ)
OxA-28969	*Cochlicella acuta*	170–180	-5.81‰	1833 ± 26	cal AD 85–235
OxA-28968	*Cochlicella acuta*	255–265	-6.53‰	2718 ± 27	910–815 cal BC
OxA-28967	*Cochlicella acuta*	288–300	-6.45‰	3187 ± 27	1500–1420 cal BC
OxA-28966	*Cochlicella acuta*	310–332	-3.52‰	3957 ± 29	2570–2350 cal BC
Beta-280906	*Prunus* sp. charcoal	332	-23.80‰	3950 ± 40	2570–2310 cal BC

There is a rich molluscan fauna in almost all 31 samples from the exposure, with a total of 20,593 shells from 26 taxa, with strong predominance of open country species at all levels; numbers of shells are higher in each of the six stabilisation horizons than in the intervening blown sand deposits. With the radiocarbon dates it is possible to correlate most of these horizons with the sediments in the main archaeological site.

The old ground surface, 310–332cm [2], lies on the head [1]. Charcoal from the interface of these deposits was dated to the late Neolithic (2570–2310 cal BC), where there is little evidence of occupation in the valley site. The molluscs here are mainly open country species, but with a good representation of catholic taxa. Of importance is the dating of *Cochlicella acuta* from this deposit; the date of 2570–2350 cal BC equates exactly to that of the charcoal from the base of the same sand, suggesting that the mollusc date is reliable. This shows that *c. acuta* was present at this site during the Neolithic. The species was previously unknown with certainty before the Bronze Age (Lewis 1968, 315) and was not thought to be well established in Cornwall until the beginning of the first millennium BC (Lewis 1975; Spencer 1975) and not until the Iron Age in the Hebrides (Evans 1971b, 59; Evans 2004). The present date pushes its appearance in the south-west of Britain back by 1500 years.

The stabilisation horizons correspond very well chronologically with those on the main archaeological settlement in the Red River valley. That at 288–300cm [4], dated here to 1500–1420 cal BC, the middle Bronze Age, correlates closely with that of the Phase 3 in the settlement, dated *c.* 1500–1200 cal BC; the mollusc numbers increase with *Vallonia excentrica* being the dominant species, implying that this area of the towans was short grazed grassland. The next period of stabilisation, 245–265cm [6], occurred in the late Bronze Age, 910–815 cal BC, around the time when the settlement in the valley (Phase 5) was beginning its decline. There is still a strong presence of *Vallonia* with greater numbers than *Pupilla muscorum*, showing that this was still grazed grassland. A thick layer of windblown sand [7] covers the late Bronze Age horizon and corresponds to the instability which probably led to the abandonment of the valley settlement.

A further stabilisation phase is present at 160–190cm [8], dated to the Roman period, 85–235 cal AD. There was no known occupation in the valley at this time, but the homestead on the cliff edge at Porth Godrevy, less than 500m to the north, is known to date to this period (Fowler 1962), and it seems likely that this area of the towans was part of their agricultural land; the molluscs show that this was still open country, but with more *Pupilla* than *Vallonia* it was unlikely to have been intensively grazed.

Further blown sand [9] overlies this horizon. The very high numbers of molluscs at 110–123cm is not related to a stabilisation horizon and is perhaps due to an inclement weather phase with winnowing of the shells. A rather insubstantial stabilisation horizon is present at 68–83cm [10]. Blown sand again accumulated [11] until stabilisation of the modern subsoil [12] and turf [13] sands. The modern rank grassland with reduced grazing is reflected in the molluscan fauna which, while still predominantly open country, contains some species requiring some shade: *Lauria cylindracea*, *Aegopinella nitidula*, *Oxychilus alliarius* and *Punctum pygmaeum*.

Strap Rocks (A) – Walker

A more detailed study of the Holocene stratigraphy in the area of Strap Rocks has been made as part of the present investigations and will be described in full. The location is 150m north-east of Musset's sample site (Strap Rocks A).

Atlantic gales regularly impact on the coasts of Cornwall and erosion of the cliffs and dunes is a constant threat. New features may be observed eroding out of newly exposed cliff sections, and one such feature was recently observed 700m south of the Red River at Gwithian. During 2011 Prof. Martin Bell noticed a line of large stones high in the cliffs under a covering of blown sand (Figure 145), towards the northern limit of Strap Rocks, between Gwithian Towans and the Red River estuary. He considered that these stones did not look natural and could be a man-made wall. He suggested that further investigation would be appropriate, particularly to determine whether this may be a prehistoric wall or associated with Industrial Age mine workings. The location (grid ref.: SW 5802.4163) is at the south end of Godrevy beach near the northern limits of a line of low cliffs which extend southwards along the coastline of Gwithian Towans (Figure 146) and rising to a height of 10m.

Figure 145. The cliffs adjacent to Strap Rocks B, showing the blown sand and lines of stones above the cliff rock. The red dot indicates the location of the mollusc column.

A visit to the site in July 2012 showed that there was a solid platform high on the cliff very close to the stone layers, approximately at the level of the presumed wall (Figure 146). This provided a sound and safe base from which to obtain a mollusc column. The land is owned by Cornwall County Council and permission to do the required work was obtained from Alison Clough, Countryside Officer for the Council. The location is just outside the SSSI of Godrevy and Hayle Towans so consent from Natural England was not necessary.

Figure 146. The stone layers in the cliff face at Strap Rocks B seen from the platform at the sampling site.

Archaeology

There have been no excavations in the immediate vicinity but a cist and walling have previously been observed eroding from the cliff at the base of the dunes (Jones 1998); the walls were considered to be parts of medieval or prehistoric field systems, and the cist may be that found in 1741 containing an urn and bones (Polsue 1868, 162). Their position in relation to the wall investigated in the present study is shown in Figure 147. There is evidence of previous human activity revealed by a mussel midden, of unknown age, seen eroding out of the sand 2m inland from the new stone line (Figure 147), only observed during a site visit in February 2014 following the exceptionally severe winter storms which had battered southern England during the previous weeks.

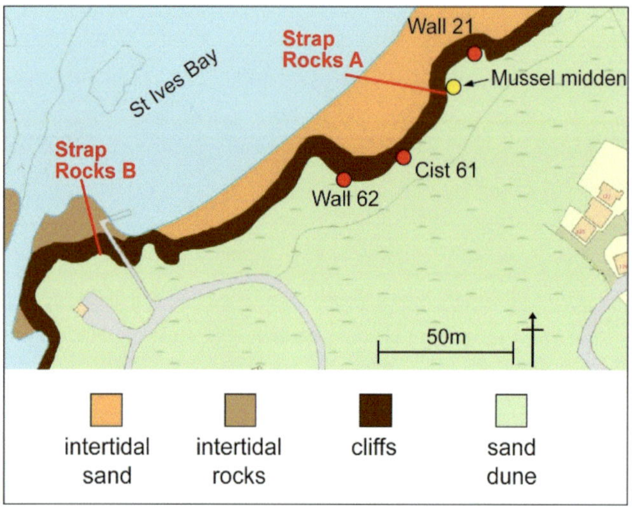

Figure 147. The Strap Rocks area showing the sites of the walls and cist eroding from the cliff and the locations of the mollusc columns. Previously recorded walls and cists from Jones (1998).

Figure 148. The mussel midden eroding from the sand at Strap Rocks.

Surface collection of artefacts has taken place here since the late 1950s over an area of about 700 x 200m, and produced material dating to the Mesolithic, Bronze Age, Romano-British, post-Roman and medieval periods (Sturgess 2004, 119, 180). A small excavation in 1963 of a midden at 'Sandy Lane' (500m east of the present site, see Figure 142) exposed by winter gales, produced quantities of medieval and post medieval pottery, but also a few sherds of Bronze Age and Iron Age/Romano-British pottery (Thomas 1964b; Sturgess 2004, 203). Thomas considered Sandy Lane to be a medieval site, perhaps the location of the settlement of Conarton mentioned in Domesday.

The abandoned mine shaft of Wheal Emily is 160m to the south-west. Details of this mine are unclear, but it produced copper, zinc and lead until it closed sometime between 1860 and 1887 (Dines 1956, 149; Hamilton Jenkin 1963, 44).

The mollusc column was obtained on 14 September 2012, with suitable precautions being taken to cordon off the area of beach below the site to preclude injury to people on the beach from falling debris. An exposed dune section was cleaned to a vertical face (Figure 149). The uppermost layer, 0–20cm, consisted largely of a dense matt of marram grass roots. Disturbance from rabbit burrowing was evident at 30–40cm, but the stratigraphy above and below this seemed undisturbed. There was yellow blown sand from the surface down to 94cm where there was an abrupt change to a pinker more silty sediment. At 105cm large stones were encountered which had a level upper surface and appeared to have been laid with abutting edges (Figure 150). They were on the same level as the stones in the cliff and are consistent with a line of walling. It was only possible to obtain samples for a further 15cm inside the stone line as it was not considered justified at this time to extend the exposure.

Figure 149. The exposure after removal of the mollusc column.

Figure 150. The layer of stones corresponding to the top of the probable wall seen in the cliff face.

A column of samples was acquired, each of about 3kg, commencing from the surface downwards. Instability of the sands precluded sampling from the base upwards, but care was taken to ensure that there was no contamination of samples with debris from higher levels.

After completion of sample acquisition the small pit was backfilled and the cut into the sand face smoothed to leave as little evidence of disturbance as possible.

Mollusc analysis

1kg samples from the mollusc column were wet sieved and the molluscs extracted and identified using standard procedures. The shells found in the samples are shown in Table 36. There was a total of 5807 non-marine shells identified from 24 taxa. A few mussel fragments were present in the 94–105cm sediments, and charcoal fragments in each of the 94–130cm samples, although very scarce in the 94–105cm deposit. Some rabbit fur was present in the 30–40cm sample entirely consistent with this level being part of a rabbit burrow.

Table 36. The molluscs found in the Strap Rock samples. 1kg samples.

Depth (cm)	**0-20**	**20-30**	**30-40**	**40-50**	**50-60**	**60-70**	**70-80**	**80-90**	**90-94**	**94-100**	**100-105**	**105-110**	**110-120**	**120-130**
Munsell number	10YR 6/4		10YR 4/3	10YR 6/3					10YR 5/4					
	turf	sand	rabbit burrow	sand						clearance			wood/scrub	
Carychium minimum		1										3	1	2
Cochlicopa lubrica	2		2	11	2	10	7	3	4	31	42	5	2	
Cochlicopa lubricella		1				3	3	1	1	16	13	5		
Cochlicopa spp.	3	2	1	14	13	19	15	8	11	59	69	39	14	2
Vertigo pygmaea	10	3	2	11	1	6	11	4	14	86	64	23	5	6
Pupilla muscorum	28	4	12	45	34	49	39	32	11	34	49	16	4	13
Lauria cylindracea	1		2	9	8	15	3	4	7	34	54	32	11	7
Acanthinula aculeata														2
Vallonia excentrica	80	23	19	152	65	137	96	44	56	335	286	142	30	25
Punctum pygmaeum	2			16	4	4	8	2	4	10	7	4		1
Discus rotundatus										2	1	14	34	34
Vitrina pellucida	3			4	2	6	5	6	3	2	1	1	1	3
Vitrea contracta														7
Vitrea crystallina				1		1		2		10	8	11	11	2
Aegopinella nitidula		1								17	23	28	24	30
Aegopinella pura														5
Limacidae						2					3	2		
Clausilia bidentata	5													
Cochlicella acuta	161	81	63	278	160	157	150	154	100	298	399	95	1	14
Ashfordia granulata	2	1	4	6	6	4	3	5	4	45	50	26		1
Cernuella virgata	44	23	4		17	2	7	1	1					
Helicella itala	120	34	11	42	18	3	10	2			3			
Xerocrassa geyeri										3	25	27	11	9
Cepaea sp.					1		1				4	1		1
Cornu aspersum		1			1									
TOTAL SHELLS	**461**	**175**	**120**	**589**	**332**	**418**	**358**	**268**	**216**	**982**	**1101**	**474**	**149**	**164**
TOTAL SPECIES	**12**	**11**	**9**	**11**	**13**	**14**	**13**	**13**	**11**	**14**	**17**	**17**	**12**	**18**
mussels										y	y			
charcoal										y	y	y	y	y
beetle					1									
rabbit fur			y											

These mollusc numbers are shown graphically in Figure 151 (absolute numbers of shells) and Figure 152 (relative numbers of shells).

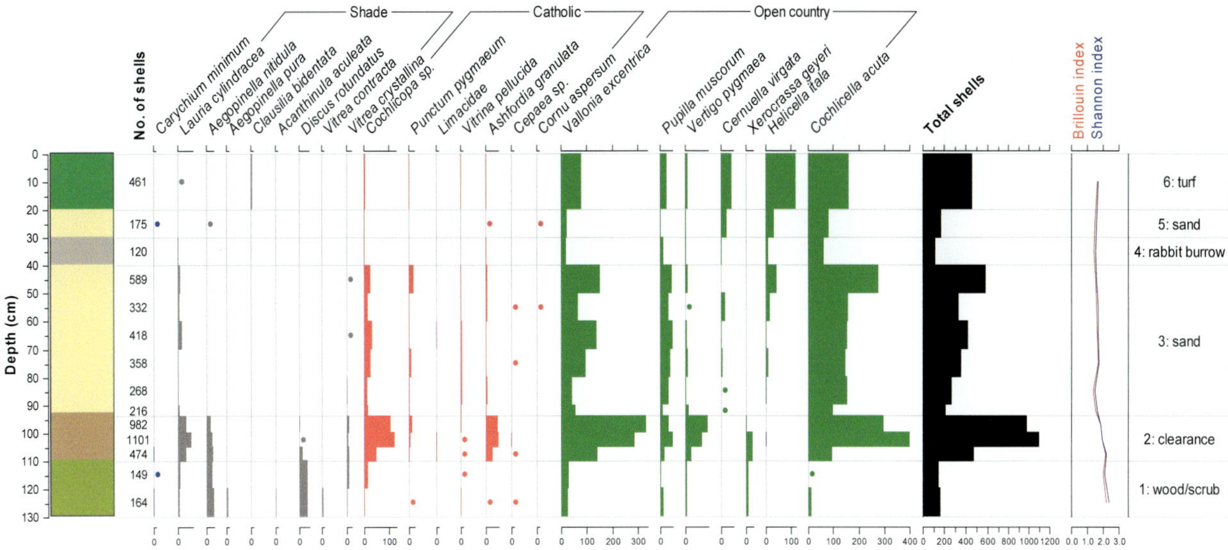

Figure 151. Mollusc diagram for Strap Rocks.
Absolute numbers of molluscs. Dots indicate single specimen in that sample.

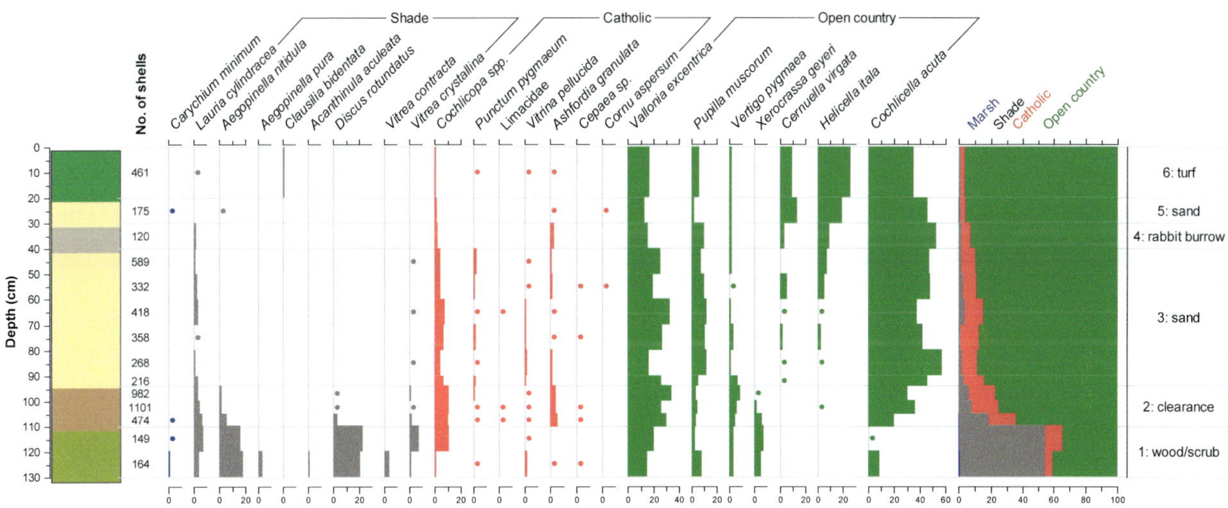

Figure 152. Mollusc diagram for Strap Rocks.
Relative numbers of molluscs. Dots indicate <1% of that species in the sample.

Other analyses

Particle size, loss on ignition analyses and mollusc numbers are summarised in Figure 153. pH was also measured but is not shown as every reading was between 8.6 and 9.2, the highest measurements being at 60–80cm. Radiocarbon dating was performed on molluscs from three samples from the lower part of the column, shown in Table 37. The selected samples were at the lowest level of blown sand (*Cochlicella acuta*), immediately above the top of the stone line (*Xerocrassa geyeri*) and the lowest level reached in the mollusc column (*Xerocrassa geyeri*).

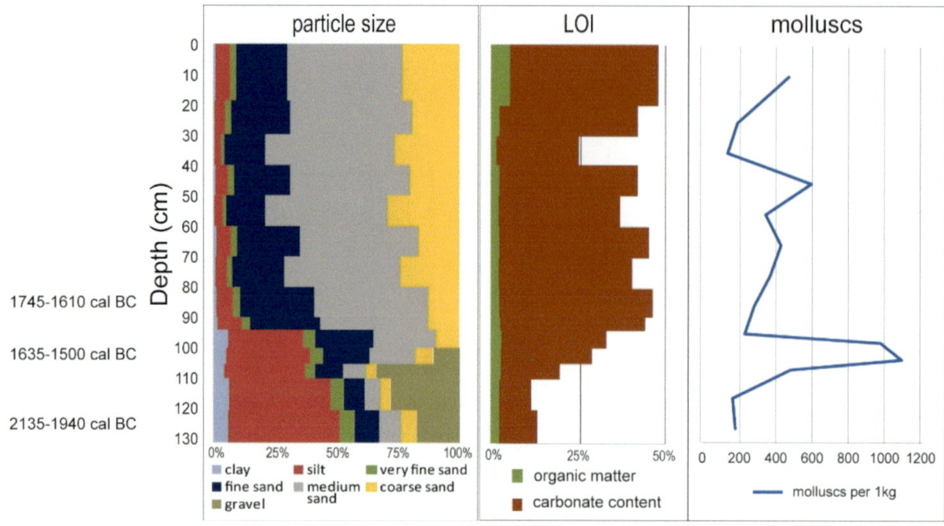

Figure 153. Summary diagrams for Strap Rocks.

Table 37. Radiocarbon dates for the Strap Rocks molluscs.

Lab number	Material	Depth	$\delta^{13}C$	Radiocarbon date (BP)	Calibrated date	Mid-date
OxA-28972	*Cochlicella acuta*	80–90	-6.40	3370 ± 29	1745–1610 cal BC	1678 BC
OxA-28971	*Xerocrassa geyeri*	100–105	-7.59	3290 ± 29	1635–1500 cal BC	1567 BC
OxA-28970	*Xerocrassa geyeri*	120–130	-7.55	3650 ±2 9	2135–1940 cal BC	2037.5 BC

Discussion of the Strap Rocks section

There is a distinct difference between the sediments above and below 94cm. Above this level there is windblown sand with typical rounded grains, whereas below this level the grains were much more angular, strongly implying that they are not of blown sand origin (Figure 154). There is also abrupt change in particle size distribution, with the lower sediments containing 38–67% clay/silt while those above contain only 4–12% clay/silt. There is no significant alteration in the organic matter content, although the carbonate content is lower in the deeper samples. The high proportion of clay/silt and the appearances of the sand grains below 94cm suggest that these lower sands derive from an old ground surface with reworked Pleistocene Head. Differences in the sand at this site have previously been noted by Rogers (1909) who stated: 'At Gwithian, on the south side of the Red River near Strap Rocks, old and modern sand may be seen close together, separated only by a thin layer of stony Head.' That layer of Head was not seen in the present section.

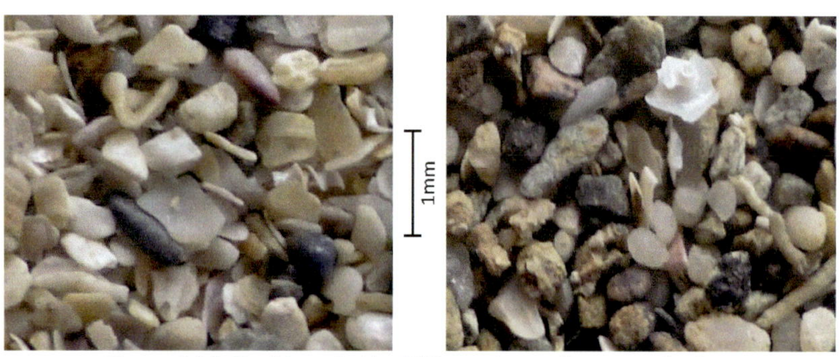

Figure 154. Residues in the 500µm sieve after wet sieving.
Left: sediment at 80–90cm with rounded sand grains.
Right: sediment from 105–110cm with angular sand grains.

The lowest two samples [1] contain a molluscan fauna that has a large proportion of species which require shade (over 50%), some needing more than is provided at the base of grass tussocks or similar vegetation. *Discus rotundatus*, *Acanthinula aculeata*, *Vitrea contracta*, *V. crystallina* and *Aegopinella pura* are all associated with woodland, often being found in leaf litter, some requiring higher degrees of moisture than others, but all capable of surviving in more open environments. *Aegopinella nitidula* also requires shade, but is regularly found at the base of rank vegetation, and wood/scrub is not necessary. However, the open country species *Vallonia excentrica* and *Xerocrassa geyeri* are also well represented, the latter being a xerophile shell of very dry open areas. This is a very mixed assemblage, with both shade and open country taxa. The mollusc assemblage indicates that there was open woodland/scrub in the vicinity of the wall. The sands of this zone are from the Pleistocene Head, either deposited against the wall as a revetment or blown/washed to this site from further inland. The date (2135–1940 cal BC) places it firmly in the early Bronze Age. The charcoal fragments support there being human presence at this time.

The sediments from 94–110cm [2] contain very high numbers of shells, over 1000 per kg. The proportion of shade species is markedly reduced, to around 7%, being replaced by those found in open country, especially *Vallonia excentrica* and *Cochlicella acuta*, the latter being very scarce in lower sediments. *Xerocrassa geyeri* is frequent in the lower part of [2] but is less common in its topmost level. *Vertigo pygmaea*, a shell especially of dry short turfed grassland, increases in number, while *Lauria cylindracea*, a species of dry shady places commonly found under rocks or stone walls, is also present. This is a mixed assemblage, although the diversity indices show it to be somewhat less diverse than the lower deposits, despite the much greater shell numbers; nevertheless, it clearly represents a stable landscape. Much of this sand covers the remaining structure of the wall and is likely to have derived from sediments that are becoming increasingly stable above the wall, probably with some grazing to keep the vegetation low, but still tall enough to provide shade for *Aegopinella nitidula*. These levels should, perhaps, be seen as a transition zone between the lower sediments and the more open country above, and is likely to represent a period of woodland/scrub clearance. The date of this horizon (1635–1500 cal BC) overlaps that of the overlying blown sand (1745–1610 cal BC), indicating that at this particular site blown sand did not start to accumulate until close to the end of the early Bronze Age. Eleven centimetres of Pleistocene sand cover the upper surface of the stones, suggesting that this field wall went out of use prior to the deposition of blown beach sand. However, the dates show that this deposit is contemporary with the base of the blown sand, and it may be that some of the stones were lost soon after burial. The wall in this study appears to be only about 50cm in height (not measured accurately as the exposure in the cliff face is not accessible), but wall no. 62 further south observed by Jones (1998, 92) was 'under 1 metre in height', so loss of wall height would seem indicated. Mussel shells with a few charcoal fragments were present in this sediment, and may link chronologically with the eroding midden illustrated in Figure 148.

Although conjectural, sand blow could have been the precipitating factor for abandonment of these coastal fields. Certainly, major sand blow is known in this immediate area in more recent times. Writing in 1868, Polsue states that:

> The barton of Upton, one of the principal farms in the parish, was overwhelmed and now lies buried in the sands. This was so suddenly done during one tempestuous night that the members of the family were obliged to make their escape through the chamber windows; shortly after the house disappeared.

The upper sediments, from the ground surface down to 94cm [3, 1] contain almost entirely open country molluscan taxa, although there is a slightly higher proportion of catholic taxa in the lower half of this sand. These are open sand dunes assemblages with areas of blown sand providing adequate shade at the base of vegetation for some of the catholic and few shade requiring species to live. Of note is the presence of five specimens of *Clausilia bidentata* in the turf sample. At present, there is no woodland or scrub in the immediate area, although marram grass is growing and provides shade. Three of the five shells in the sample are juveniles with translucent upper whorls, while the other two are more mature apical fragments. Whether they are autochthonous or have been deposited here by wind or animal action is open to question, but there is probably sufficient shade for them to have lived at this site. It is also important to note that there are specimens of *Cernuella virgata* in the majority of the samples above 94cm, but no *Xerocrassa geyeri*. *Helicella itala*, an obligatory heliophile typical of dry grassland, becomes increasingly common towards the surface and is common only in the turf sample.

The rabbit burrow level, 30–40cm [2], contains similar molluscs to the sands above and below it, although with somewhat reduced numbers of both shells and taxa. It is clear that the burrow has become infilled with sand from the vicinity.

There is sufficient evidence to conclude that this structure is a field wall of early Bronze Age date, and it is therefore likely that the other walls along this stretch of coast are of a similar date. This is an important result because it indicates the presence of Bronze Age activity in the towans south of the Red River in addition to those previously investigated on the main Gwithian site north of the river. The lower portion of the deposits, below 94cm, are likely to be a buried land surface covered by colluvial sediments which built up against the wall. Bronze Age lynchets marking field boundaries were previously shown at the main Gwithian site (Megaw 1976) and the present investigation at Strap Rocks suggests a similar landscape. The walls were constructed at a time when the environment had evidence of woodland or scrub, prior to the deposition of the first blown sand, and the clearance of the wood/scrub predated the initial sand blow episodes. Evidence of former woodland conditions on Cornish coastal blown sand sites are few. While Spencer (1975) implied such woodland at Gwithian, it is only at Newquay that there is strong evidence of woodland prior to clearance or burial by sand (Kennard and Warren 1903; Woodward 1908; Spencer 1975; Milles 1991); away from the coast there is more evidence, for example at Daymer Bay (Arkell 1943; Milles 1991) and Harlyn Bay (Bullen 1902; Bullen 1912; Whimster 1977).

The date of wall construction has not been established, but it continued in use until around 1600 cal BC before being overwhelmed by sand blow which seems likely to be the cause of its abandonment.

Strap Rocks (B) – Musset

A mollusc column was obtained in 2010 by Simon Musset from an exposure opposite Strap Rocks (SW 57887.41568) (Figure 155), to the south of Godrevy Bay, and south of the Red River (Musset 2012). The Holocene dune sands at this location overlie Head which in turn overlies a raised beach and a Quaternary wave-cut rock platform. The molluscs from the column have all been re-examined by the present author. The findings are shown in Figure 156; the single *Cornu aspersum* found in deep sediments is thought to be intrusive.

Figure 155. The position of Musset's mollusc column at Strap Rocks; the buried soil is seen at the base of the exposure. (photo: Martin Bell)

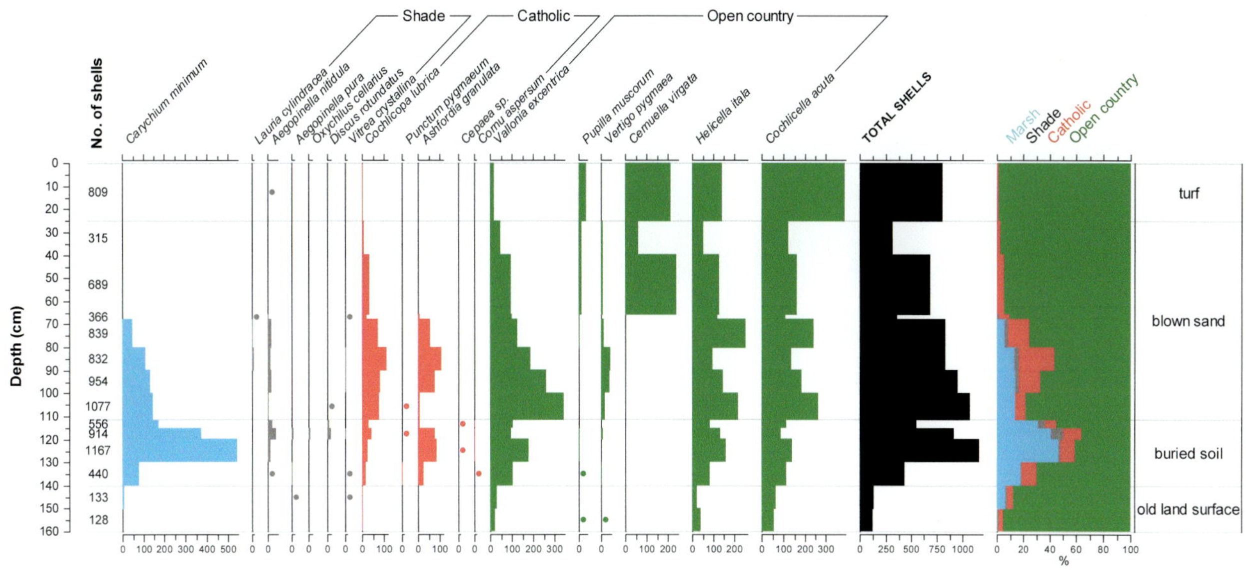

Figure 156. Mollusc diagram (absolute numbers) for the mollusc column obtained by Musset at Strap Rocks; mollusc identifications are by the present author.

The major difference in this column when compared to other sites is the presence of large numbers of *Carychium minimum* in the mid and lower levels, particularly in the buried soils from 112–140cm. This taxon, while occurring in leaf litter, generally prefers damp grassland such as marshes. Although the location is now at the shore edge and being steadily eroded by winter storms, it seems probable that when these sands were deposited they were well inland where marshland could develop. This supposition is strongly supported by an investigation by James (2008) of the Quaternary rock section at this precise location; he found that the 'unique presence of dip sections through the exposures at Strap Rocks is interpreted as a fossil storm beach barrier which could have impounded a former marsh in front of the low fossil cliff.'

As the marshland began to dry with accumulation of blown sand open country molluscs come to dominate, but with persistence of shade and catholic species; the land was probably grazed as reflected in the numbers of *Vallonia excentrica*, but grazing declines over time with development of longer grassland, although with minimal appearance of any shade species.

Sandcot – Kevill

In 2011 a mollusc column was obtained by Richard Kevill (2012) from an exposure at Sandcot cottage, east of the Bronze Age settlement, during excavations for a house extension at Sandcot (SW 59192.42277) (Figure 157). Only some of the samples were studied at the time, but all the samples have been analysed and re-evaluated by the present author (Figure 158). No dating was obtained but some correlation with the chronology of the Bronze Age site is possible. The old land surface contained numerous open country molluscs; the dominance of *Vallonia excentrica* implies short grazed grassland, and there is no good evidence of significant shade. This is overlain by 150cm of blown sand which did not have any visible stratification, but the molluscs show two zones with high numbers of shells, implying some degree of stabilisation, which probably correspond to the Bronze Age levels in the main settlement site. Although open country taxa dominate, there are good proportions of shade and catholic species, increasing upwards from 200cm to 90cm, being maximal at 100–120cm. The presence of Zonitidae and *Discus rotundatus* strongly imply that there is shade in the vicinity, probably scrub woodland such as is found on the towans at the present time which probably correspond to the middle Bronze Age level described by Davies (2006; 2007, 64). A stabilisation horizon (identified by being much darker in colour and more compact than the sand above and below) from 45–83cm represents a buried soil showing a grassland fauna, without any scrub, and it is likely that the cultivated area of the settlement extended this far east, either during the late Bronze Age or in the post-Roman period. This soil is covered by blown sand with few shells indicating instability, perhaps with considerable winnowing, which extends up to the modern turf containing a mixed fauna typical of long rank grassland.

Figure 157. The exposure at Sandcot, showing the buried soil horizon between two deposits of blown sand. (photo: Martin Bell)

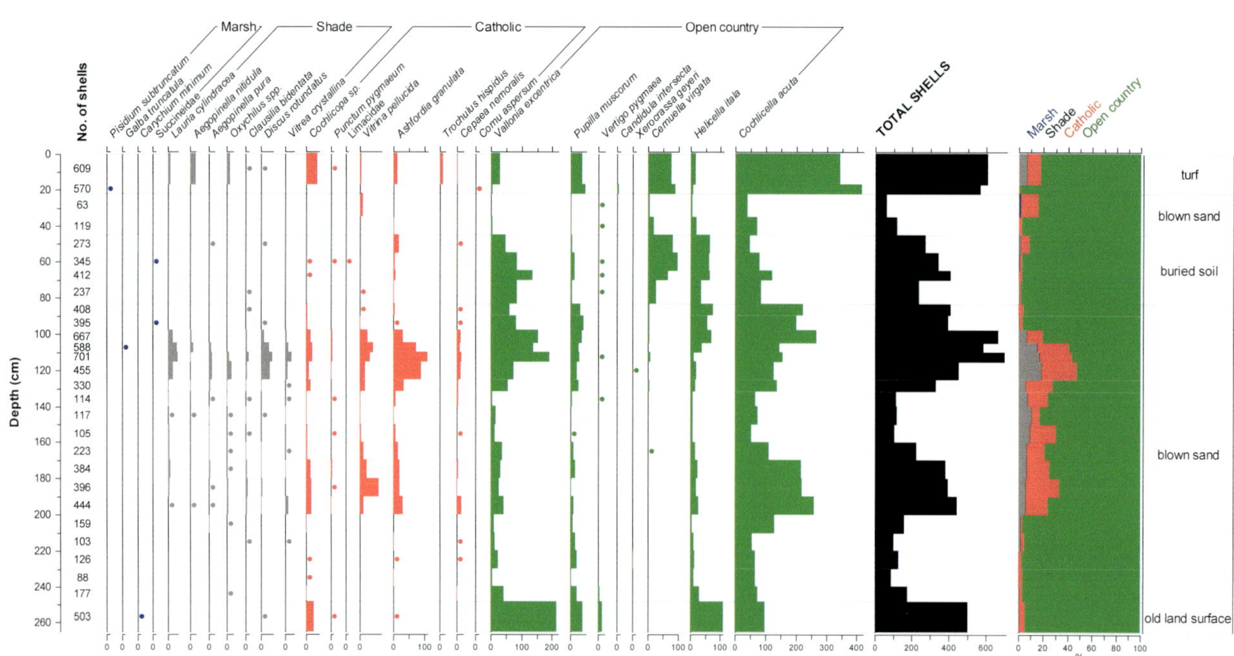

Figure 158. Mollusc diagram (absolute numbers) for the mollusc column obtained by Kevill at Sandcot; mollusc identifications are by the present author.

Valley Floor – Bull

Kerrie Bull (2011) examined a transect of cores taken in the flood plain north of the Red River, to the south-east of the post-Roman industrial site (SW 58953.42094 to SW 58987.42065). Although eight cores were obtained only three were analysed in detail (20m, 25m, 35m); unfortunately, the samples are no longer available for further study. Figure 159 shows the profiles obtained from the cores, when five units (zones) were identified. Stabilisation horizons were present at deeper levels in all cores. The basal unit 5 considered to be estuarine alluvium, consisted mainly of gleyed sand with few land shells, which were mostly open country taxa although with some catholic species. Numerous mussel fragments and a few other small marine shells were present as well as rounded stones in the cores at 20m, 25m and 35m, strongly supporting the presence of a midden covering at least 15m in extent; its depth suggests that it may be early Bronze Age.

Unit 4 represents stabilisation horizons with the molluscs showing an approximately equal proportion of open country and catholic taxa and occasional shade species (although none in the 35m core); this unit may correspond to the Bronze Age levels on the settlement site, when the ground was rank grassland permitting a wide diversity of molluscan fauna to thrive. A few wet ground species were present (*Galba truncatula*, Succineidae, *Vertigo antivertigo*) in the 20m core consistent with being close to the line of the river. Units 3 and 2 (the difference being that unit 3 was mainly gleyed) consisted of a deep accumulation of blown sand and contained very few shells in any core, entirely open country taxa (except for two *Cepaea* sp.). The modern turf, unit 1, was almost devoid of molluscs, those few present being open country species.

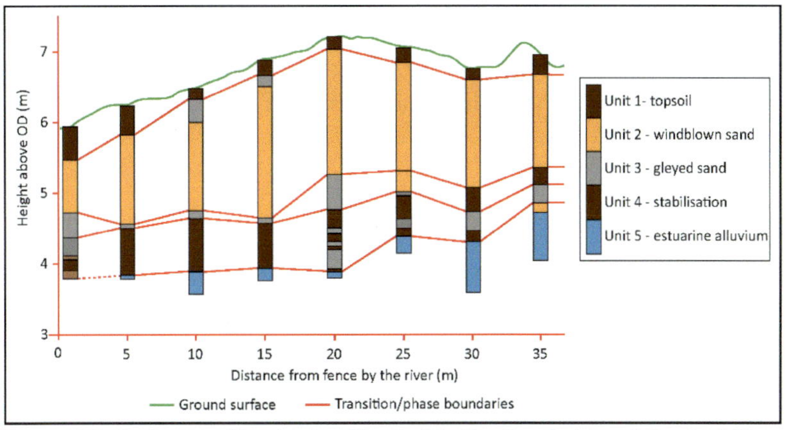

Figure 159. Coring profile obtained by Bull in the flood plain of the Red River.
(OD levels determined from subsequent measurement of TBM with Leica differential GPS). (redrawn from Bull 2011)

These cores, obtained about 300m east of the transect line described earlier, show similar profiles to the more superficial sediments found in the 0.5m and 15m transect cores. They did not reach sufficient depth to determine whether the marsh levels seen in the transect extended this far up the estuary.

Chapter XIII – Mining for tin and other metals

Mining for tin and other minerals has been carried out in Cornwall for many centuries, although this ceased following the closure in 1998 of South Crofty mine at Camborne. The importance of the mining industry to the history of Cornwall was recognised when the Cornish Mining World Heritage Site was established in 2006 incorporating mining areas in many parts of the county.

The earliest written records of specific mines within the Red River catchment date from the 1580s (Hamilton Jenkin 1965, 5) but it is likely that knowledge of Cornish tin extends well into prehistoric times. 'Tin Islands' were mentioned by Hecataeus of Miletus (*c.* 549–476 BC) on his map of the world (unfortunately no longer surviving; a reconstruction is shown in Figure 160) who depicted them lying off the coast of north-west Europe. Pytheas of Massilia (Marseilles) circumnavigated Britain in the fourth century BC and implied a tin trade between Cornwall and western France (Cunliffe 2008, 347).

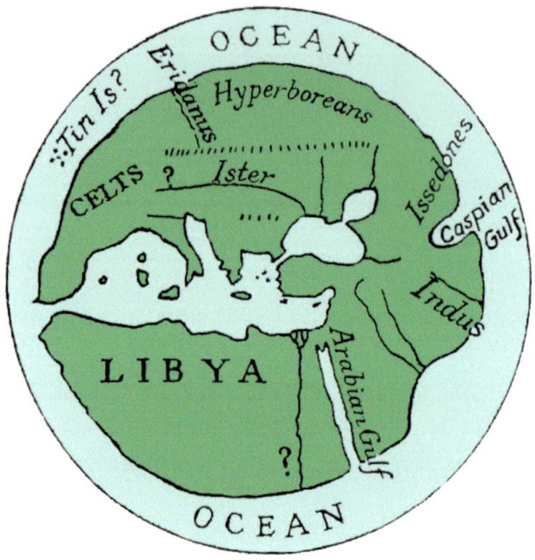

Figure 160. Reconstruction of the Map of the World by Hecataeus of Miletus. (after Thomson 1965, Fig. 11)

Texts by Strabo, who lived from 63/62 BC to about AD 24, are among the earliest surviving written references specifically to Britain as a source of tin:

> The westerly parts of Britain lie opposite these headlands towards the north; and in like manner the islands called Cassiterides, situated in the open sea approximately in the latitude of Britain, lie opposite to, and north of, the Artabrians.
>
> (Strabo *Geographica*, 2.5.15; translation: Jones 1917) .

> [Tin] is produced both in the country of the barbarians who live beyond Lusitania, and in the Cassiterides Islands; and tin is brought to Massilia from the British Islands also.
>
> (Strabo *Geographica*, 3.2.9; translation: Jones 1923) .

> The Cassiterides are ten in number, and they lie near each other in the high sea to the north of the port of the Atrabrians. One of them is desert, but the rest are inhabited by people who wear black cloaks, go clad in tunics that reach to their feet, wear belts around their breasts, walk around with canes, and resemble the goddesses of vengeance in tragedies. They live off their herds, leading for the most part a nomadic life. As they have mines of tin and lead, they give these metals and the hides from their cattle to the sea-traders in exchange for pottery, salt and copper utensils. ... After Publius Crassus crossed over to these people and saw that the metals were being dug from only a slight depth, and that the men there were peaceable, he forthwith laid abundant information before all who wished to traffic over this sea, albeit a wider sea than that which separates Britain from the continent.
>
> (Strabo *Geographica*, 3.5.11; translation: Jones 1923) .

Strabo's original maps no longer survive but a reconstruction from his texts indicate where he thought the Casseritides may lie (Figure 161), showing these islands north of Spain, where no such islands exist.

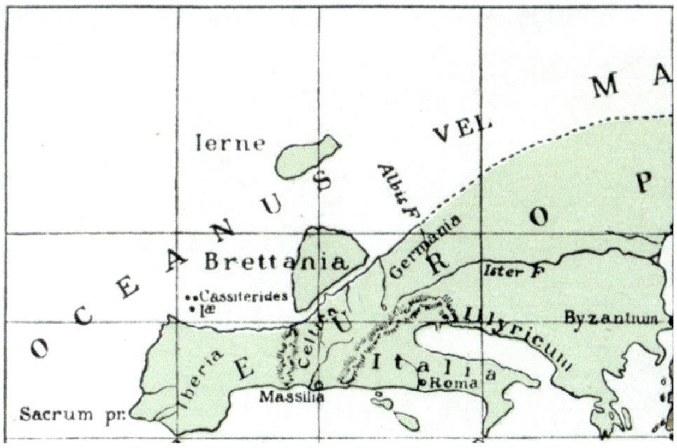

Figure 161. Europe according to Strabo, showing the Cassiterides Islands north of Iberia. (portion of frontispiece from Jones 1917)

The ore from which tin is extracted is cassiterite (SnO_2), and the location of the 'Cassiterides' or 'Tin Islands' is unclear, with many islands or groups of islands being debated, although *The Oxford Classical Dictionary* (Hornblower and Spawforth 2005) states:

> The unambiguous evidence about the location of the Cassiterides in the classical sources suggests that it was a partly mythologized generic name for the sources of tin beyond the Mediterranean world and not a single place. The absence of archaeological evidence for any pre-Roman iron age trade with the tin-producing areas of SW England supports this.

Diodorus Siculus, a Greek from Sicily, writing around the end of the first century BC, described the tin trade on the mainland of Britain. His 'Belerium' is considered to be the Land's End peninsula; the identity of Ictis is a matter of much debate, although either St. Michael's Mount or Mount Batten near Plymouth are likely (Cunliffe 2005, 472).

> Now we shall speak something of the tin that is dug and gotten there. They that inhabit the British promontory of Belerium, by reason of their converse with merchants, are more civilized and courteous to strangers than the rest are. These are the people that make the tin, which with a great deal of care and labour they dig out of the ground; and that being rocky, the metal is mixed with some veins of earth, out of which they melt the metal, and then refine it; then they beat it into four-square pieces like to a dye, and carry it to a British isle near at hand, called Ictis. For at low tide, all being dry between them and the island, they convey over in carts abundance of tin in the mean time. But there is one thing peculiar to these islands which lie between Britain and Europe: for at full sea, they appear to be islands, but at low water for a long way, they look like so many peninsulas. Hence the merchants transport the tin they buy to the inhabitants to France; and for thirty days journey, they carry it in packs upon horses' backs through France, to the mouth of the river Rhone.
>
> (Siculus *The historical library of Diodorus the Sicilian* 5.2; translation Booth 1814: 310).

When tin was first obtained in the west country is unknown, although it is probable that ore was originally available in alluvial gravels on Dartmoor during the middle/late Bronze Age (Gerrard 2000, 17; Thorndycraft *et al.* 2004). Slag and artefacts associated with archaeological sites of a similar period have been recorded (Penhallurick 1986, 151; Penhallurick 1997). Trace element studies have suggested prehistoric metal mining (or perhaps widespread burning) in the early Bronze Age (*c.* 2500 BC) and middle Bronze Age (*c.* 1600 BC) (Webster 2007, 87). There is less evidence in the late Bronze Age (Buckley 2005, 14) but tin smelting was found in the Iron Age villages of Chysauster, Goldherring and the Roundago of Castallack (Gerrard 2000, 19). The finding of a socketed axe mould at Gwithian, thought to date to the 12th–11th centuries BC (see Chapter II) does, however, strongly imply that there was metal working in the catchment area of the Red River during the late Bronze Age.

There is good archaeological evidence for tin mining in the Roman-British period (Buckley 2005, 18), but no written records survive. The first documentary evidence dates from 1156 (Lewes 1906, 524) when Pipe Rolls of Henry II refer to mining in Devon and Cornwall. During later medieval times the tin trade was of sufficient importance in Devon and Cornwall that Stannary Courts were established to administer the law independently of normal legal processes (Penhallurick 1986, 148). It was not until the Industrial Revolution and the development of pumping engines to remove water that exploitation of deep tin mining became possible. Large areas of Cornwall were subsequently given over to deep mining.

Other metals are also available in Cornwall, principally copper, lead, zinc, tungsten and arsenic. There is, however, no evidence of prehistoric copper mining in Cornwall (Sharpe 1997; Timberlake 2009, 100). The extent to which minerals containing other metals were obtained during the Bronze Age has not been determined. The early Bronze Age Nebra Sky Disc from central Germany is composed of different types of gold, some of which has been shown to come from Cornwall, but the copper was shown not to be Cornish in origin (Ehser *et al.* 2011).

THE MINING PROCESS

Minerals may be extracted either by lode mining, when ores are obtained directly from the rocks in which they are formed, or from tin streaming, the process of recovering tin from mine tailings or natural washings at or downstream from the original ore sites.

Convincing evidence for prehistoric lode mining in Cornwall is hard to come by, as the region has been so exploited over the last 4000 years that signs of early quarrying may have been lost. However, substantial centres of early Bronze Age lode mining have been found in Wales, especially at The Great Orme (Dutton and Fasham 1994) and Cwmystwyth (Timberlake 2003). Penhallurick (1997) considers a middle Bronze Age palstave illustrated by Borlase (1872, 41) to come from lode mining. Further evidence comes from the fact that there is no alluvial copper in Cornwall, so any copper mined must have come from lode working (Penhallurick 1997). Underground lode mining became the principle method of extraction by the end of the fifteenth century (Buckley 2009, 125). The development of pumping machines in the nineteenth century to remove water and to transport miners from the ground to the working areas enabled ever deeper seams to become accessible (Richardson 1974, 64). Great depths were reached, the deepest being at Dolcoath mine (Figure 162) at Tuckingmill, Camborne where 925m was achieved in 1910 (Buckley 2010, 362).

Figure 162. Dolcoath mine in the foreground, looking east towards Redruth, 1892/93.
Each chimney represents a separate mine working. (© The Cornish Studies Centre, Redruth: ref. corn00352)

Tin streaming refers to the process of washing cassiterite ore once it has been extracted from the ground, and was originally synonymous with opencast mining (Penhallurick 1997); it is only later that the term became more associated with the recovery of tin lost into the river system from upstream mines. Prehistoric artefacts have been recovered from many tin streamworks in Cornwall, but none from the Red River catchment (Penhallurick 1997, Figure 16). Streaming probably became established around the late fifteenth century when lode mining began to produce large quantities of tailings, and continued into the nineteenth and twentieth centuries along many Cornish rivers.

Tin streaming in the current sense consists of using gravity to separate the heavy tin ores from lighter clays and sand found in alluvial deposits. This is in contrast to lode mining when the rock has to be crushed and ground to concentrate the ore and extract the minerals. No native alluvial tin has been found in the Red River catchment area (Penhallurick 1997), but large quantities of waste material were discharged into local river systems. Quantification of the mineral losses from mines is difficult, but it is clear that substantial amounts of tin and other minerals failed to be extracted at the mines. In 1890, for example, 66 Cornish mines sold 12,944 tons of black tin for £700,962, while 1,730 tons worth £69,206 were recovered from Red River streaming – 10% of the total value (Thomas 1913); Thomas estimated that 'the average loss in dressing tin in Cornish mines at present must certainly exceed 33%.' It is important to appreciate that the process of load mining results in the build-up of sediment which leads to the formation of dams or lakes unless it is flushed, when a proportion settles out as sediment in steams and river. In contrast, tin streaming is the process of removal of sediment from river beds.

THE RED RIVER AND ITS CATCHMENT AREA

One of the most developed mining regions in Cornwall was around the towns of Redruth and Camborne, an area which at one time was the richest tin mining district in southwest England (Dines 1956, 276; Yim 1981). This is the catchment area of the Red River, which arises south-east of the modern town of Camborne and flows north between Camborne and Redruth before turning west towards Gwithian and which drains into the east side of St. Ives Bay (Figure 163). Many of the mines in the Camborne area were immensely profitable; an area north of the Carn Brea range of hills, covering only about 650ha, had yielded an aggregate amount of £2 million by 1871 (Thomas 1871), an enormous sum for those times.

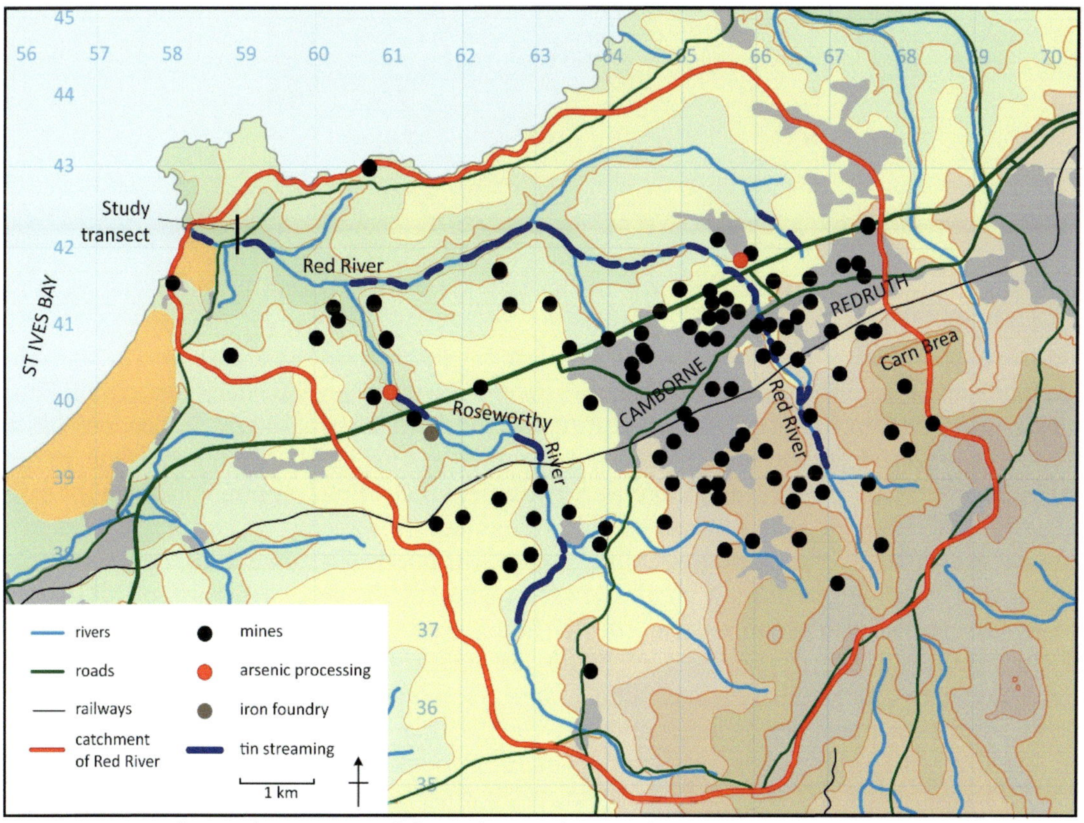

Figure 163. The catchment area of the Red River showing the mines and tin streaming locations. The map also shows the location of two arsenic processing plants and an iron foundry. The present study transect is marked near the outlet of the Red River into St Ives Bay.

With its tributaries it was the main tin streaming river in Cornwall, with over 40 separate activities in the nineteenth century (Barton 1969, 152, 182). Although originally named the Conor Dour it earned its new name of 'Red' from the colouring caused by mine effluent, mainly soluble ferrous (Fe^{2+}) and ferric (Fe^{3+}) compounds and insoluble ferric hydroxide ($Fe(OH)_3$) which forms red/orange 'ochre' at low pH (Younger 2002) and which increased following the expansion of lode mining. Since the closure of the last mine in the catchment area in 1998 (South Crofty) there has been a programme to clean the waters, and the river currently flows clear. However, in the past its name was clearly justified. Figure 164 shows its colour close to one of the mine adits, while Figure 165 shows its exit to the sea at Gwithian, giving a very good indication of the spread of colouration into St. Ives Bay.

Figure 164. The adit of Dolcoath mine into the Red River. (photo: Plymouth University)

Figure 165. The Red River outflow in July 1978. (photo: Ian Stehbens)

The numerous mines in the catchment area of the Red River in Cornwall produced waste which was discharged into the river system. It is difficult to determine exactly how many mines were active at any one time due to regular name changes and amalgamation of pits. Figure 163 shows nearly 100 mines but doubtless many more have left little or no trace and there were often several shafts for an individual mine. In the majority of these tin, in the form of cassiterite, was the main mineral sought, but copper was another major mineral extracted from this area of Cornwall. Arsenic was important and obtained from mines in the Redruth/Camborne area (Earl 1996) and there were two arsenic calcining (refining) plants in the catchment area (marked in red on Figure 163), each depositing tailings into the streams. Arsenic is a toxic element, and its deposition in sediments beside the river during periods of flooding may have affected the wellbeing of both animal and plant life, especially in the lower reaches of the river where it passes through the Reskajeage marshes below the confluence of the Red and Roseworthy Rivers, and on the flood plain north of Gwithian village. However, arsenical compounds are poorly taken up by plants during growth and quantities found in modern foods grown on alluvial silts are unlikely to be seriously harmful to humans or livestock (Xu and Thornton 1985; Mitchell and Barr 1995). Other elements mined commercially included tungsten and zinc, with small quantities of lead, silver, cobalt, nickel and bismuth.

The first documented lode mine in the Red River catchment dates from 1584 when John Norden mentioned tinworks at Tolcarne (Hamilton Jenkin 1965, 5). It would seem probable that quarrying for copper and tin predates this and tin streaming almost certainly existed along the Conor Dour, as the Red River was then known, during medieval times. Geochemical analysis of the sediments accumulating in the river valley may provide some clues as to the timing. Some minerals sink to the river bottom during their passage to the sea and become attached to particulate matter. Yim (1981) studied fluvial bottom sediments at eleven sites along the Red River and two in its tributary, the Roseworthy River, and found high concentrations of multiple metals along the rivers. He found that there was high positive correlation of levels of tin with those of iron, tungsten and arsenic in these bottom sediments and demonstrated that residues of tin, copper, zinc, iron, arsenic and lead in the river were the result of mining activity and not natural erosion processes. Hosking and Ong (1963–64) looked at tin and other heavy metals deposited on the Gwithian/Hayle beach on the west side of St Ives Bay and concluded that most of the heavy elements derived from mine tailings in the Red River catchment, although some zinc and lead came from local cliff lodes. Similar findings are described from the Camel and Gannel estuaries in the north Cornish coast (Pirrie *et al.* 2000) and in the Fal estuary in south Cornwall (Pirrie *et al.* 2002; Pirrie *et al.* 2003).

Tin streaming has already been noted as being prolific in the Red River catchment. As refining processes became more efficient the importance of streaming declined. However, sufficient quantities of tin reached St Ives Bay that a new streamworks was constructed between the Gwithian road bridge and the sea in the mid-nineteenth century to recover tin from beach shingle and sand in Godrevy Bay. The Gwithian Streamworks continued in operation until the beginning of World War II (Kirkham 2005, 19). However, that may not be the end of the story, as new consideration is being given to recover tin from the sea bed in St Ives Bay (BBC Cornwall News Report, 7 February 2013).

GEOCHEMISTRY OF THE RED RIVER VALLEY CORE

The history of mining, particularly for tin, in the catchment area of the Red River suggested that analysis of cores taken from the valley could provide insight into the chronology of mining and metal working in the area. Although much of the river bed has been disturbed by stream mining there is no historical evidence that such disturbance has ever taken place at the location where the cores in the present study were obtained. The sequences and chronologies obtained for this project show that they are not disturbed by mining activity.

The chemical composition of sediments was studied in the 0.5m core by fine resolution Itrax XRF scanning (measurements at 1mm increments). Further details of the technique are given in Chapter IV. It is emphasised that the Itrax method measures radiation counts per unit time and not element concentration. The magnetic susceptibility of the 0.5m core was also measured at 5mm increments with the Itrax analyser.

One of the problems with Itrax analyses is that the emitted signals from calcium and tin are very similar, and it is not possible to separate the two elements. This is unfortunate in that tin, in the form of cassiterite, was the principle ore sought in the mining process. A wide range of other metals can, however, act as proxies for mining activity. The main tin ore, cassiterite, is closely associated with wolframite and arsenopyrite (Dines 1956, 20; Bishop *et al.* 1999, 50), so tungsten and arsenic levels may provide a guide to tin extraction; both of these metals were mined commercially in the nineteenth and twentieth centuries (Dines 1956, 23, 25). Sphalerite, the main mineral containing zinc (Dines 1956, 26; Bishop *et al.* 1999, 24), was also mined in the Camborne area and is found in association with lead and sometimes copper, all of which may provide a guide to mining activity. Other metals ascribed to mine tailings contaminating the Red River are iron, manganese and yttrium (Yim 1981; Morris *et al.* 1985).

A further problem is the degree to which elements and minerals are retained in the sediments deposited in the valley floor. When the sediment consists mainly of sand-sized grains then there is little retention of minerals, giving misleading results. In the figures that follow the periods when sand was the dominant deposit have been highlighted.

The Itrax XRF graphs for the 0.5m core are shown in Figure 166. The chronology shown on the Y-axis is derived from the radiocarbon and OSL dating on the 30m core, as discussed in Chapter IX. All the graphs show +ve or –ve variation from the average measurement before the commencement of the Bronze Age (2200 BC), prior to any likely mining activity in the area. It is important to point out that the dates on these figures are extrapolated as carefully as possible from the 30m core, but there will inevitability be some discrepancies with true dates.

The periods of 1310–400 BC and AD 1690–1850 are those where measurements are affected by the presence of significant quantities of sand, and do not give a true reflection of the chemical composition of sediment; these periods will not be discussed further.

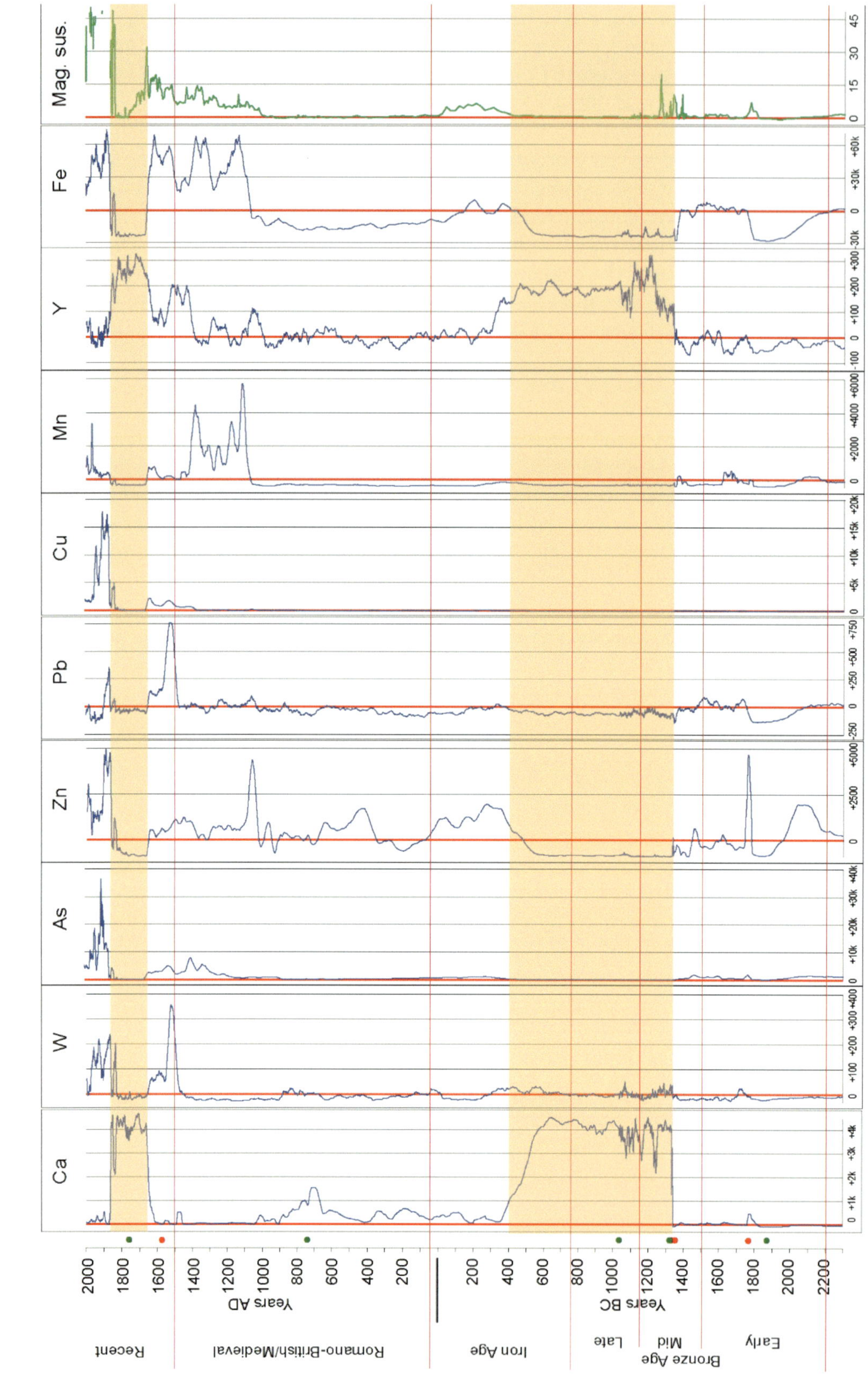

Figure 166. Itrax XRF geochemistry and magnetic susceptibility on the 0.5m core.
Values are differences (- or +) from the average background counts/10 seconds prior to the Bronze Age (2200 BC) with the baseline indicated by the red line on each graph. The highlighted areas are those where the sediment is sandy with high calcium levels.
The dots indicate dated levels, extrapolated from the 30m core.

Peaks of different metals/minerals are evident at different times. The first detected rise is about 1750 BC, when there is slight increase in tungsten, arsenic and lead, marked increase in zinc and some elevation of magnetic susceptibility. These increases are short-lived, reducing to background levels after only a few years. The OSL date range of 2070–1650 BC is within the early Bronze Age, but is this sufficient evidence to suggest that there was mineral extraction activity during this period? If mining did take place, then it is probable that cassiterite, the main tin ore, was the mineral being sought, but that metal cannot be detected by the analytical methods used. The small increase in tungsten and arsenic levels is consistent with the presence of wolframite and arsenopyrite, ores associated with cassiterite. The high level of zinc, from sphalerite, is another pointer to mining at this time. The lead signal is likely to be incidental. These metal levels are not strong enough to provide solid evidence of mineral extraction during the early Bronze Age, although there are tentative indications that this did occur.

During the early part of the middle Bronze Age none of the measurements show any variation from background levels. Unfortunately, sand levels after 1320 BC preclude any interpretation of mining during the later Bronze Age, although there is some rise in magnetic susceptibility which may imply the presence of haematite. The lack of mineral signals is regrettable, as fragments of two stone axe moulds were found at Gwithian during the twentieth century excavations and which have been dated to the century or so before 1150 cal BC (Burgess 1976).

Some increase in zinc and magnetic susceptibility is present in the later Iron Age, but the dating at this time is not secure and no conclusions concerning mining can be drawn. The chemistry does not reveal any mining activity during the Romano-British period. There is, however, considerable elevation of several metal signals around AD 1050–1100, reflected in zinc, lead, manganese, yttrium and iron levels, but not of tungsten or arsenic. This implies that tin, in the form of cassiterite, was not being mined at this time, but that zinc minerals as sphalerite were obtained, with manganese and yttrium being incidental contaminants. While zinc may not have been utilised during this period it may be that lead-containing galena was sought, the two minerals resembling each other – sphalerite is also known as 'false galena' and 'mock lead ore'. It is suggested that the very marked increase in iron at this time probably caused the waters of the Conor Dour to turn red, and the name 'Red River' could well have been adopted for the first time.

It is not until AD 1300 that arsenic and yttrium levels began to rise, together with progressive increase in magnetic susceptibility. Increase in tungsten and lead followed about AD 1500. The earliest elevation of a copper signal is at this time, there having been no earlier rise. Manganese levels, however, fall. These findings strongly indicate mining activity during the late medieval period and early Industrial Age, and are consistent with documented mining in the Red River catchment area at this time (Hamilton Jenkin 1965, 5).

No information is gained for the period from 1650 to about 1870 due to lack of metal signals associated with the presence of sand. In the later nineteenth and the twentieth centuries there is very clear increase in all the measured metals. These findings are in accord with the widespread introduction of new technology, especially pumping machines (manufactured on a very large scale at nearby Hayle), during the nineteenth century, making the Camborne/Redruth area one of the most productive mining areas in the country.

Chapter XIV – Discussion and conclusions

THE GWITHIAN ENVIRONMENT

The Gwithian coring study and trench excavation, together with the mollusc columns at Strap Rocks and Godrevy Towans, add to knowledge concerning the landscape of the lower Red River valley during the Holocene. The new dates, together with those obtained from the 1950s to 1960s excavations and the 2005 trench, allow a sediment sequence at Gwithian to be proposed (Table 38).

Table 38. Chronology of the Gwithian palaeoenvironment. The settlement activities did not necessary span the full length of the time periods shown in column 1. The colours indicate differing sedimentary deposits.

Date	Valley basin	Valley sides	Godrevy Towans (west end)	Strap Rocks	Settlement activity
AD 1950–present	stabilisation	stabilisation		stabilisation	
AD 1650–1950	blown sand	blown sand		blown sand	
c AD 800–1650	marsh	blown sand		blown sand	Crane Godrevy
c AD 400–*c* 800	marsh	stabilisation	blown sand	blown sand	Post-Roman site
c 800 BC–*c* AD 400	marsh	blown sand	stabilisation	blown sand	Porth Godrevy
900–*c* 800 BC	stabilisation	stabilisation	stabilisation	blown sand	Phase 5 Gwithian settlement
1100–900 BC	blown sand	blown sand	blown sand	blown sand	Phase 5 Gwithian settlement
1150–1100 BC	stabilisation	stabilisation	blown sand	blown sand	Phase 5 Gwithian settlement
1350–1150 BC	blown sand	stabilisation	blown sand	blown sand	Phases 3+4 Gwithian settlement
c 1700–1350 BC	fen/carr woodland	stabilisation	stabilisation	buried land surface	
c 1800–*c* 1700 BC	fen/carr woodland	blown sand	blown sand	buried land surface	Phase 1 Gwithian settlement
3000–*c* 1800 BC	fen/carr woodland	blown sand	stabilisation	buried land surface	
4000–3000 BC	fen/carr woodland	initial sand blow			
pre–4000 BC					Mesolithic activity

This chronology can be suggested graphically (Figure 167). This interpretation is based on the findings in the present study together with those of previous research and may be a good indication of the environment in the close vicinity of the study areas, but elsewhere the interpretation is speculative. It has been assumed that the coastal margin was further into St Ives Bay during the prehistoric period and has been eroded to its present position over several thousand years. The projection has generally followed the modern contours in proposing areas that were subject to sand inundation, but clearly this may be a false assumption. The extent of the occupation sites has been represented to include areas of fields and cultivation where known.

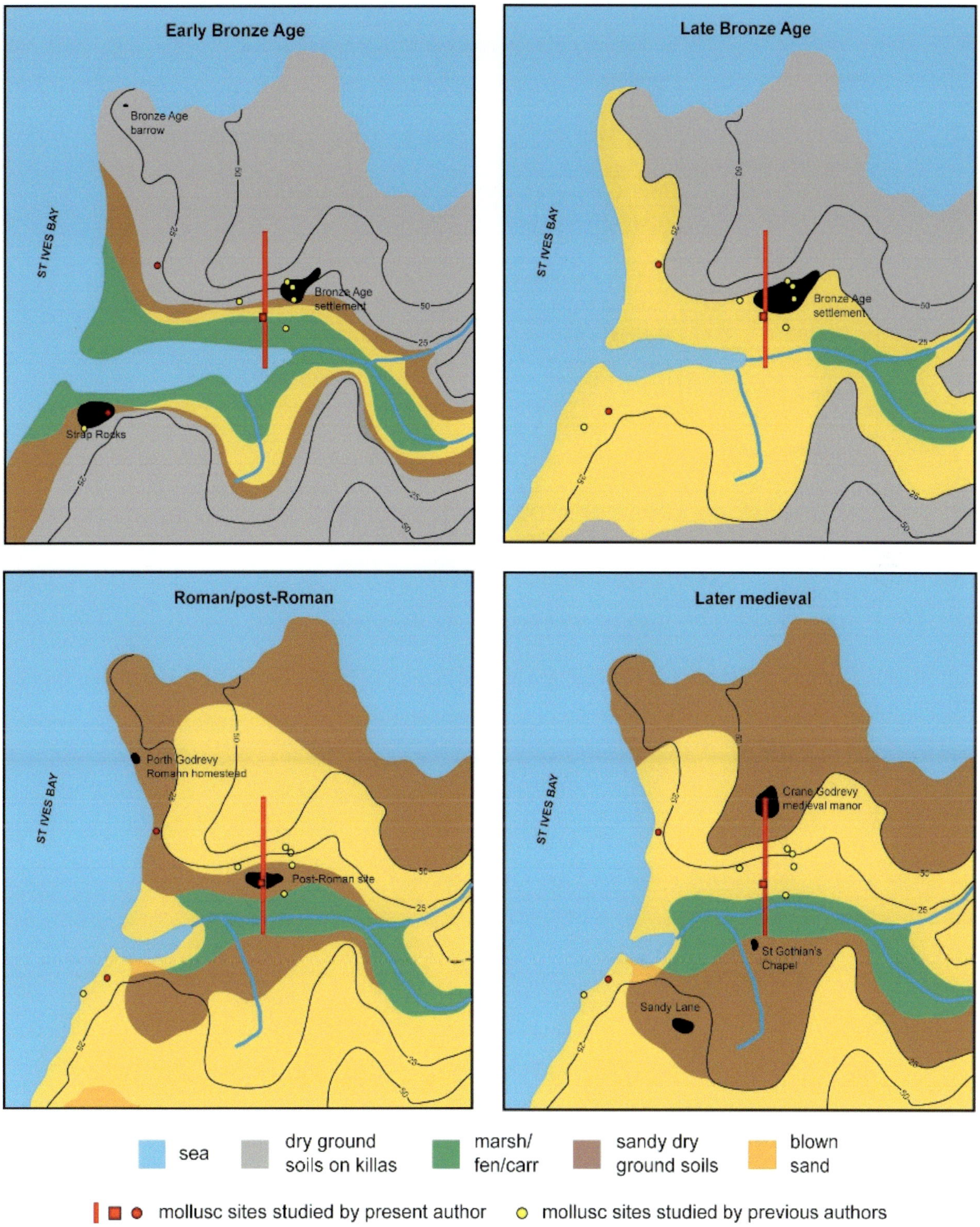

Figure 167. The landscape of the Gwithian area at different chronological periods.

Within this chronology, the specific episodes of aeolian activity can be placed in a somewhat tighter time frame (Table 39).

Table 39. Summary of sand blow events at the study sites.

Date	Location	Event
Medieval		
end 16th century AD	Crane Godrevy	abandonment, probably related to sand blow
Romano-British		
4th century AD	Porth Godrevy farmstead	abandoned due to sand blow
after AD 89–235	Godrevy Towans (west)	sand blow
Late Bronze Age		
after 913–812 BC	Godrevy Towans (west)	sand blow
after 900 BC	Gwithian Trench	sand blow
after 900 BC	Gwithian settlement	abandonment, probably related to sand blow
1220–880 BC	Gwithian Trench	blown sand
Middle Bronze Age		
1540–1120 BC	Gwithian valley	buried by sand
Early Bronze Age		
1745–1611 BC	Strap Rocks	blown sand
1800–1500 BC	Gwithian settlement	blown sand horizon
2070–1650 BC	Gwithian 30m core	sand blow
3350–2870 BC	Gwithian Trench	initial sand blow
Late Neolithic		
2572–2307 BC	Godrevy Towans (west)	initial sand blow

Figure 168 is a summary diagram of the changes at Gwithian since 3000 BC, showing correlation between settlement activity with climate episodes.

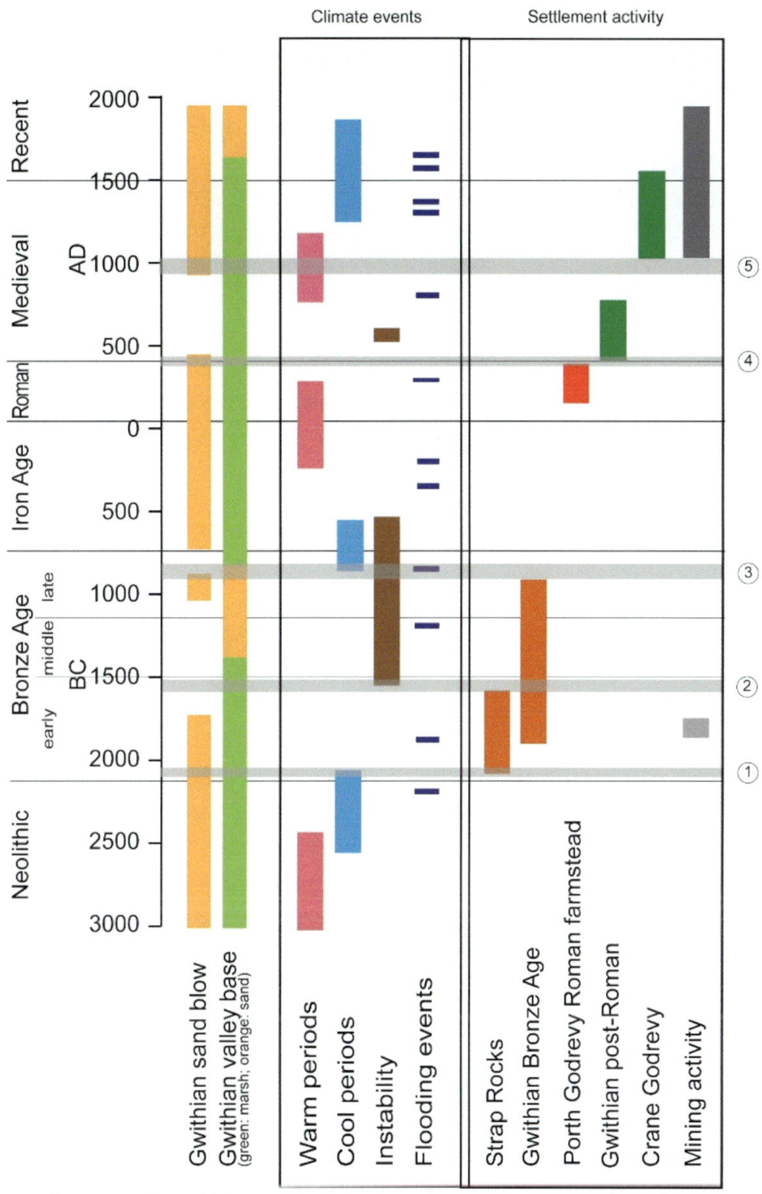

Figure 168. Summary diagram of sand blow, temperature, climate events and settlement activity at Gwithian. The numbers on the right refer to correlations discussed in the text. (Warm periods: Johnsen *et al.* 2001; Mann *et al.* 2009; Wang *et al.* 2012) (Instability: Bond *et al.* 1997; Bond *et al.* 2001) (Cool periods: Barber *et al.* 1994; Van Geel *et al.* 1996; Barber *et al.* 2000; Fagan 2000; Barber *et al.* 2003; Mann *et al.* 2009; PAGES 2k Consortium 2013) (Flooding events: Johnstone *et al.* 2006; Macklin *et al.* 2006)

Mesolithic and Neolithic

This study had not provided any information concerning the environment during the Mesolithic period, but it is likely that the lower Red River valley was wet ground during the entire Mesolithic and Neolithic periods. Alder wood in the lowest level of sediment (949–977cm) in the 30m core is radiocarbon dated to 3945–3705 cal BC, the early Neolithic. Molluscs have not survived at these depths, but pollen analysis indicates a fen/carr woodland with relatively open areas. Mesolithic flints have previously been found on many sites in the Gwithian/Godrevy area but for the first time they have now been discovered close to the later Bronze Age settlement.

The earliest evidence of sand deposition away from the valley basin is the old ground surface in the 2012 trench, which was dated by OSL to 3350–2870 BC, the middle/late Neolithic. Molluscs were few, mainly *Xerocrassa geyeri*, but sufficient to suggest that by this time the landscape was open. At the west end of Godrevy Towans there is a slightly later date of 2570–2310 cal BC from both charcoal and molluscs. These dates are in broad agreement for the first blown sand with that of 3000–4000 BC proposed by Carter (1988,

303) for the earliest dunes in Britain and with that of 3000 BC for St Ives Bay in Cornwall suggested by Lewis (1992). The open landscape supports the findings by Spencer (1975) of the loss of woodland during the Neolithic; she additionally showed that the base layers at Towan Head, , Newquay, contained shade species of molluscs indicating scrub or light woodland prior to sand deposition, but does consider that sand may have contributed to clearance rather than ascribing a purely anthropogenic cause, a suggestion that would also be applicable to Gwithian. However, given the concentration of Mesolithic activity reported at Gwithian by Thomas (1958) and Roberts(1987), it is probable that Mesolithic communities contributed to clearance as seems to have been the case in many other coastal areas of Britain (Bell 2007).

At Strap Rocks the landscape was open woodland/scrub at least until the end of the Neolithic and probably into the early Bronze Age, when field walls were constructed. Blown sand did not arrive at this site until later, after clearance, making an anthropogenic cause for loss of woodland likely.

Early Bronze Age

The earliest (Phase 1) settlement at Gwithian, excavated in the 1950s and 1960s, dates to the early Bronze Age: 1890–1610 cal BC (3430±50 BP; OxA-14568) (Hamilton *et al.* 2008). The molluscs in the lowest sand above the old ground surface in the 2012 trench were dated to 2015–1775 cal BC, in agreement with the settlement evidence, but there is a 'hiatus' in the trench of around 600–800 years, with the sand above the old ground surface dated by OSL to 1290–910 BC. Any sand that did accumulate previously in this location may have been deflated by wind.

Closer to the river the first dated sand deposition occurred about 1800 cal BC, but this only affected the valley sides, evident in the 30m core where there is a 30cm thick sand horizon between two organic silt layers, dated by OSL to 2080–1660 BC, i.e. during the early Bronze Age, at the time that the sand comprising the old ground surface in the trench was deposited. At the present river location the stratigraphy indicates continuity of marsh throughout this period. The molluscan evidence shows that the area of the meadow was still very open, but pollen indicates that the lower parts of the valley remained little changed since earlier times, being wet ground although some areas of woodland are still present. The initial building in the settlement was constructed of wood, and charcoal included oak, hazel, hawthorn, gorse and broom consistent with woodland and scrub (Nowakowski 2009, 119). It is significant that, contrary to deterministic expectation, this first settlement phase is associated not with stabilisation but with a period of sand deposition, at least in part of the landscape.

At Strap Rocks, 1.2km south-west of the main Gwithian Bronze Age site, the stones eroding out of the cliff are shown to be a wall line, presumably a field boundary. The sand against the wall, reworked Pleistocene Head, is dated to the early Bronze Age (2135–1940 cal BC at the base and 1635–1500 cal BC at the top), slightly earlier than Phase 1 of the main Gwithian settlement. This basal date correlates with amelioration of the cooler climate (Figure 168: ①). It is likely to be associated with other sections of wall which have previously been observed in the cliff face. Whether a midden close to the wall is coeval has not been established but it could suggest the presence of a nearby settlement, and a watch should be maintained in this area as erosion proceeds. The presence of the cist observed in the eighteenth century in the cliff face is another pointer to a second settlement in the Gwithian area. It must also be remembered that the sea shore would probably have been many hundreds of metres further into St Ives Bay during the Bronze Age, and that an associated settlement may have been seawards of the wall and now lost to coastal erosion. A further pointer to some settlement in the area is the finding of 21 sherds of pottery, considered to date to the Bronze Age, found in a primarily early medieval midden at the 'Sandy Lane' site, east of Strap Rocks, towards Gwithian village (Sturgess 2004, 203). Sand commenced accumulating over the field system during the later part of the early Bronze Age (*c* 1600 cal BC) and probably led to the abandonment of the fields, certainly in this area. This is approximately at the commencement of a long period of climatic instability (Figure 168: ②) and the two may well be related.

It is possible to suggest reasons why these settlements were established close to the outflow of the Red River. St Ives Bay is the most westerly sandy bay on the south side of the Bristol Channel and the first easily accessible beach for sea craft that rounded Cape Cornwall on the Land's End peninsula. Plentiful supplies of fresh water could be obtained and there was presumably access to food supplies. This site close to the bay enabled inhabitants to obtain both maritime and terrestrial resources. Its benefits for crop growing (frost free, higher temperatures and earlier growth), as well as the availability of sea weed and shell sand to improve the soils, would have made the area attractive.

The excavations on the main Gwithian site indicated that there was a period of neglect (Phase 2) with a blown sand horizon sometime between *c.* 1800 and *c.* 1500 cal BC (Sturgess 2007a), although cultivation continued (Nowakowski 2009, 120). The fields and probably the marsh at Strap Rocks were buried by sand during the later part of the early Bronze Age and may be coeval with this neglect. The timing could relate to a possible climatic downturn marked by a major flooding event recorded in Britain by Johnstone *et al.* (2006) in 1880 cal BC but is too early for the period of climatic instability outlined by Bond *et al.* (1997; 2001) of 1550–550 BC.

Middle/late Bronze Age

Despite the onset of general climatic instability, conditions within the Gwithian valley were sufficiently stable for the settlement to become re-established (Phase 3) during the middle Bronze Age (*c.* 1500–1200 cal BC) with evidence for field systems and ploughing covering about 1.5ha (Sturgess 2007a; Nowakowski 2009, 120). A stabilisation horizon with open country molluscs is identified at the western end of the Towans dated to 1500–1420 cal BC (Section 3.6). But the local landscape changed abruptly in the middle of the mid Bronze Age. The marsh across the valley was covered by blown sand. Radiocarbon dating at the surface of the marsh is 1420–1280 cal BC, while OSL dating of the base of the overlying sand is 1550–1130 BC. If the mid-point date of the various measurements is taken, then this change in the landscape occurred around 1350 BC. Whether or not there was a major climatic event leading to massive sand blow is not resolved, but the deposit, at the 30m core site, of 200cm of sand over 250 years which contains very few molluscs suggests rapid accumulation. Significantly it is now evident that the deposition of these large quantities of sand over the valley basin did not lead to abandonment of the settlement and is probably coeval with the blown sand deposit west of the main north-south field wall of the settlement described as Phase 4 (Sturgess 2007a, 29).

A thin stabilisation horizon containing very high numbers of molluscs (2616 shells per 500g) is present at the 30m core position, dated to about 1100 BC. The presence of abundant *Lauria cylindracea*, given its frequent association with stone walls (e.g. at Strap Rocks) and the known presence of walls elsewhere at Gwithian suggests that there may have been field walls close to the river, which is shown in the 15m core to be close by. Abundant wet ground molluscs show that the area was still very damp and the walls may have been constructed to demarcate the margin of the marsh.

Conditions remained relatively stable into the late Bronze Age, with short periods when soils had sufficient time to accumulate along the river banks, although still with intermittent flooding, shown by the presence of wet ground molluscs and horizons of gleyed sand. This is Phase 3-5 of the settlement, when ground was cleared and a new farmstead constructed with major field boundaries. The trench revealed plough marks and animal hoof prints indicating agricultural activity in the area. Extensive areas containing plough marks were excavated in the original excavations in the late 1950s and 1960s (Megaw *et al.* 1960–61) and were thought to be limited to Phase 3 (middle Bronze Age), but later understanding extended this to Phase 5 (middle/late Bronze Age) (Sturgess 2007a); the present study considerably increases the area of valley terrace that was cultivated during the middle Bronze Age. In the 2012 trench the lower buried soil horizon on the old ground surface dates to this period, with an OSL date of 1290–910 BC. Some bovid hoof prints are present on the surface of this sand indicating that animal husbandry extended at least this far from the settlement, not surprising due to the nature of the ground. The mollusc assemblage in this sand is comprised almost entirely of *Xerocrassa geyeri*, consistent with very open ground, probably with sparse vegetation. Some marine shells, mussels and limpets probably derive from midden spread and support manuring on the fields. Manuring with midden material was also demonstrated in the original excavations (Megaw *et al.* 1960–61).

Late Bronze Age/early Iron Age

The 2012 trench at Gwithian shows an episode of sand blow dated by OSL to 1220–880 BC; moderate numbers of molluscs indicate that it accumulated relatively slowly. This is overlain by a buried soil horizon dated to

940–640 BC and containing large numbers of open country molluscs with no evidence of shade. A stabilisation horizon at the western end of the towans is coeval, dated to 910–815 cal BC (Section 3.6), implying that this was a time of stability across the landscape, despite it being a period of North Atlantic instability (Bond *et al.* 1997; Bond *et al.* 2001). This period at the settlement site (Phase 5) was very active, being a 'farmstead' with several buildings, and evidence of both arable and pasture land. Nowakowski (2011, 122) comments that the 'settlement at Gwithian, by this phase, appears to have reached a level of comfortable maturity and we may imagine that life in the sands was, on the whole, pretty good.'

The upper buried soil horizon in the 2012 trench contained a rich open country mollusc assemblage with little evidence of shade. High numbers of *Vallonia excentrica* indicate grazed grassland, a conclusion strongly supported by the presence of large numbers of bovid hoof prints, with some of ovi-caprids, impressed into the upper surface of this sand, together with some ard-marks. The use of seaweed as fertiliser is suggested during this phase by the finding of a specimen of *Patella pellucida* on the surface of the soil, a shell which is only found in seaweed holdfasts. The dates of the trench sediments (940–640 BC) overlap with those of the final years of the settlement which was abandoned about 900 BC (Sturgess 2007a; Nowakowski 2011, 123) and it seems probable that the two are coeval.

The demise of the settlement seems to have been planned rather than an abrupt abandonment (Sturgess 2007a; Nowakowski 2011, 123). Some of the buildings were destroyed and the ruins sealed by spreads of settlement waste (Nowakowski 2011, 123). The trench, however, does indicate that blown sand at some time covered the stabilisation horizon corresponding to the phase of cooler, wetter, conditions from 800–400 BC proposed by Van Geel *et al.* (1996) and a flooding episode demonstrated by Johnstone *et al.* (2006) (Figure 168: ③). A thin layer containing sand rock fragments covers this horizon; their origin is unclear. They may have come from field walls or other structures up-slope of the trench location; wind and/or water deflated the sand deposits associated with them, leaving the heavier sandrock fragments in this 5cm thick layer. The timing of this event has not been resolved but may be related to the abandonment of the area for several hundred years.

Iron Age to post-Roman/early medieval

There is scant evidence of Iron Age presence at Gwithian. A few pottery sherds found on the Godrevy hillside and in the area of Gwithian village cover a period from the late Bronze Age to the Roman period, but no settlement or agricultural activity is known in the valley (Quinnell 2007b). The earliest evidence of occupation at Crane Godrevy on the towans summit is a round dating to the Roman period, or possibly late Iron Age, and could indicate that agricultural activity moved from the Bronze Age site in the valley to the higher ground of the towans.

Dating in the valley after the Bronze Age is insecure as it is not possible to determine whether the later sands accumulated steadily or intermittently, and whether quantities of sand were lost to deflation, either by wind or by water from the river. Conditions in the valley certainly changed during this period, with the return of marsh on the north side of the valley, but not on the south which remained sandy, although intermittent gleying of the sand suggests that flooding for extended periods of time did occur. No specific cause for this is revealed in the study, but there are several major flood episodes recorded in lowland Britain (Johnstone *et al.* 2006) during this period, 320 cal BC, 200 cal BC, cal AD 300, implying stormy conditions and these could have affected Cornwall, although the overall temperature remained stable. The earliest of these flood events fits chronologically with the loss of sand and flooding of the valley leading to a prolonged period of marsh, an environment that persists to the present day at Reskajeage Marshes, which commence only 200m upstream from the study site. Access to St Ives Bay may have been difficult which may be why the Roman homestead on Godrevy Headland is situated on high ground well above the Bay, although it would still have been liable to adverse storms from the west. A stabilisation horizon is present in the Godrevy Towans scarp section, dated to cal AD 85–235, fitting well with the proposed date of the homestead occupation situated 500m to the north.

The homestead was abandoned probably in the fifth century AD, approximately at the time when the post-Roman industrial site on the dune in Gwithian meadow was first established, and when there was a period of relative stability from sand blow (Figure 168: ④). The only evidence in the present study of human activity during this period is the 260cm thick midden deposits at the 30m core location; these were almost certainly connected to the post-Roman industrial site on the sand dune to the north-east. An OSL date in the middle of this deposit is AD 650–810, agreeing closely with the period of occupation determined from previous studies (Sturgess 2007b; Hamilton *et al.* 2008). This midden was localised in extent, not being found in the cores on either side.

The marsh conditions varied considerably at different times and places. At 0.5m the pH dropped as low as 2 during the early accumulation of sediment, but then rose sufficiently to permit a flourishing assemblage of wet ground molluscs to survive; nearby, at 15m, there is an even richer wet ground mollusc community at the equivalent level to that at 0.5m. Pollen confirms the increase in herbaceous wetland, but some open woodland/scrub remains, with suggestion of an anthropogenically modified landscape. This points to a spatially varied environment.

There was no indication in the trench of any buried soil or agricultural activity during the post-Roman period which would correlate with that found in the adjacent field system to the north of the trench (Fowler and Thomas 1962). Whether this area was never part of the post-Roman system, or whether evidence of activity has been lost to deflation, cannot be determined.

Medieval and recent

The northern side of the Red River valley remained marsh at least until the end of the medieval period, following which sand again covered the whole valley floor, corresponding to a period of reduced temperature and widely attested increased storminess during the Little Ice Age consistent with climatic instability. At the trench site there was continual accumulation of sand, with an OSL date of AD 1670–1730 at 15cm above the buried soil horizon. Whether this 15cm represents the total depth of sand deposited over several hundred years or there was intermittent deposition with or without deflation is not known.

The study reveals evidence of more extensive human activity than previously recognised high on Godrevy Towans during the medieval period. The medieval manor house of Crane Godrevy, dating from the late tenth or early eleventh century, is located on the summit of the towans, and the present study, while away from the site of earlier excavations, has expanded the area of probable activity on the surrounding hillside. This date approximates to further sand blow episodes in the Gwithian valley, although not sufficient to cover the valley base with sand (Figure 168: ⑤). This also correlates with the commencement of mining activity in the Red River catchment area, but whether the two events are connected is not known.

At the southern limit of the flat Towans plateau, at 495m, very high numbers of *Lauria cylindracea* suggested that there were field walls close to the core location. This supposition is supported by the finding of *Cecilioides acicula*, given the frequent association of this species with former agricultural land. This burrowing shell was found to a depth of 55cm and could derive from higher levels, but its association with the *Lauria* increases the likelihood of field systems extending this far from the manor. A probable ditch was revealed by ground penetrating radar at 530m, some 50m south-west of the closest ditches excavated previously (Thomas 1969; Sturgess and Lawson Jones 2006b). The site was abandoned around the end of the sixteenth century. No dates have been obtained from the sand that covers the buildings, but it is highly likely that sand blow may have played its part, possibly by deflating what little fertile sandy soils there may have been on the towans summit. Also, the extensive and thick sand layers downslope of Crane Godrevy (over 3m) and at the west end of the towans may have inundated the fields associated with the medieval manor house.

The steep scarp of Godrevy Towans has, until recently, been a continuously open environment with little molluscan evidence of more than short vegetation. Although much of this slope is in modern times covered in scrub, there is nothing to suggest that a similar habitat existed earlier. At 289m, about one-third of the way up the scarp, deep sand (in excess of 340cm) accumulated on a hillside devoid of tall vegetation throughout its time of deposition and provided sufficient sand for quarrying, as shown on a map of 1850–1851 (Figure 13).

TIDAL ESTUARY AND SEA LEVEL

Data obtained from the sedimentology, pollen and diatom analyses show that the northern side of the lower Red River valley adjacent to the excavated settlements was never a tidal estuary, contrary to previous opinion (Thomas 1958; Nowakowski *et al.* 2007). The cores on the south side of the valley failed to reach solid geology and marine incursion in that area is therefore not excluded. By combining the present findings with the depth of marine clay found in the boreholes from the 1980s and 2002, when clay was found at depths varying from 1.20m to 8.40m, it is possible to reconstruct approximate outlines of the extent of the marine estuary which probably extended a short distance inland from the present shore line (Figure 169). The earliest 'prehistoric' estuary penetrated on the south side of the valley no more than 1km inland from the modern shoreline. The extent of tidal influence then retreated, and two phases can be determined: 'intermediate' and 'late'. Although no dates can be ascribed to these phases they are probably related to rising sea levels. The suggested position of the 'late' inlet corresponds to the course of the Red River prior to nineteenth century rerouting of the river (see Figure 13). There is no evidence at any of the locations sampled in the present study of the gravel and clay horizons described by Stephens (1899, 1899–1900) (see Figure 14); either he misinterpreted his findings (unfortunately his detailed reports do not seem to have survived), or there was a narrow, winding, river channel which has not been discovered in the present study.

The earliest dated sediments at Gwithian are the organic silts laid down on the underlying killas on the north side of the valley at 3945–2705 cal BC, when there is no evidence of blown sand nor of any marine sediments. This, together with the other measured dates can be plotted on published sea level curves for south-west England (Figure 170). It is stressed that these are the maximum possible heights for sea level, and the actual level could have been lower. These points from the Gwithian study fit very well with the previous curves, being closer to those from elsewhere in the Bristol Channel than that from the English Channel. There are no previous studies with data from the north Cornish coast for comparison.

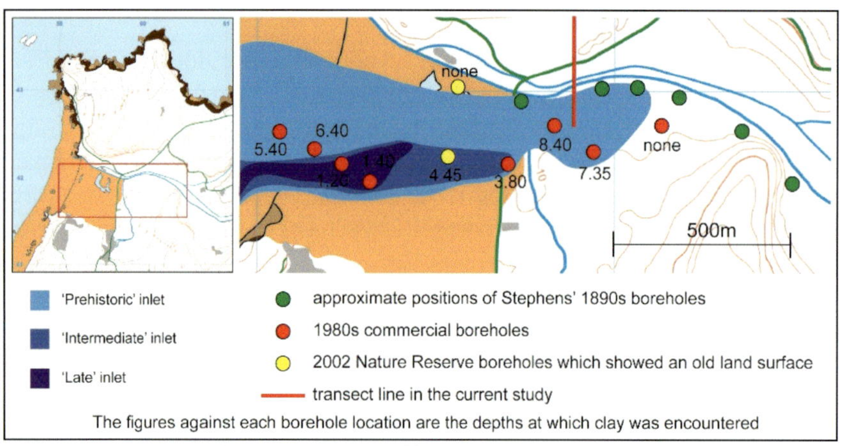

Figure 169. Suggested tidal extent at different times in the lower Red River valley.

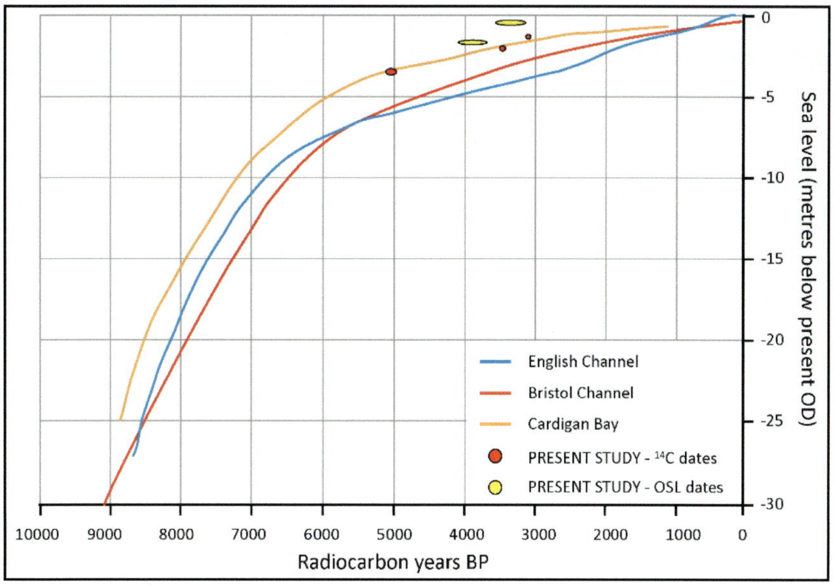

Figure 170. Sea level curves for south-west England and Wales to include the results from the present study. Note that the time scale is in radiocarbon years BP, so uncalibrated radiocarbon dates as well as the OSL dates have been used for this graph.
The dates for the present study are those when there is known to be dry land, and therefore the highest possible sea level. (after Heyworth and Kidson 1982)

MINING IN THE RED RIVER CATCHMENT

Geochemical analyses have allowed some insight into the chronology of mining activity in the Camborne area, the tailings of which flow with the Red and Roseworthy Rivers into St Ives Bay at Gwithian. Although it was not possible with the XRF methods used to evaluate levels of tin due to its similarity to calcium signals, other metals and minerals can act as proxies.

During the early Bronze Age there is a short period, around 1750 BC, when there is a marked increase in zinc levels and small increases in tungsten, arsenic and lead and with some rise in magnetic susceptibility. These changes are possible indications of mining activity upstream, but are not conclusive. The lack of any copper signal is not surprising, there being no evidence elsewhere of copper being extracted in Cornwall during the Bronze Age (Sharpe 1997; Timberlake 2009, 100). It is unfortunate that this study cannot provide evidence of any mining activity during the second half of the Bronze Age, as minerals only adhere readily to fine grained silt or clay particles and not coarse sand grains. The sediments at the sample site were sandy from *c.* 1350 BC to *c.* 400 BC.

There is no evidence of mining residues during the post-Roman period. This might seem unexpected in view of the industrial nature of the buildings on the nearby dune (Sturgess 2007b), where metal residues suggest that smithing was undertaken (Hatton 2004). However, it is considered (Thorpe and Thomas 2008) that the source of iron may have been from concretions obtained from the local cliffs at Gwithian (Thomas 1957–58).

The first good evidence of mine contamination in the Red River sediments is seen about AD 1050–1100. Initially there is a marked rise in zinc consistent with sphalerite extraction, and of manganese and iron, with small rises in lead and yttrium. The strong increase in iron concentrations and in magnetic susceptibility is also good evidence that mine tailings were entering the river system; the river waters must have turned red at this time, and the river name subsequently became known as the 'Red River'. It is not until AD 1300 that a slow rise in arsenic levels is detected followed by large increases in lead and tungsten at AD 1500, almost certainly indicating extraction of cassiterite for tin; copper only shows a small increased signal at this latter date, having been only minimally evident for the previous 300 years. These findings clearly indicate that mining was active in the catchment area during the late medieval and early Industrial Age, as is known from the written records (Hamilton Jenkin 1962).

How do these results correlate with the sedimentology of the cores? At 0.5m there was fall in acidity to pH 2, probably during the post-Roman period, but the XRF does not reveal any evidence of mining activity at this time. It is therefore more likely that the period of extreme acidity was due to stagnant ponding of water in the area of this core rather than mine contamination. However, at 15m, a period of acidity (to pH 3.8) occurred around AD 1100–1450 which correlates very well with the evidence given in the previous paragraph, when there was considerable elevation of the signals of several metals. It is therefore suggested that mining contributed to these changes in the local environment.

Sand precludes analysis from AD 1650–1870, but after this there is dramatic increase in most metal signals, entirely consistent with the expansion of mining activity during the nineteenth century. It is only very recently, towards the end of the twentieth century, that tungsten, arsenic, copper and iron signals show reduction towards baseline levels contemporary with the cessation of mining. South Crofty, Europe's last tin mine, finally closed in 1998 and all mining activity around Camborne and Redruth ceased.

It has been shown that the upper 50–60cm of the sands in the fields south of the Red River and those close to the north side of the river are virtually devoid of molluscs. Some of the uppermost layers south of the river have been removed to construct a windbreak but the remaining sand seems unexpectedly sterile. It is known that flooding from high water levels occurs during periods of adverse weather such as during the winters of 2013 and 2014 (Figure 171), although overbank flooding is not a regular occurrence (A. Thomas, pers. com.). It seems reasonable to suspect that some of the mining contaminants affected the fields during these periods of high water leading to the relative sterility of the present findings.

Figure 171. Flooding in the lower Red River valley, February 2014.
View looking south across the valley from Godrevy Towans.

MOLLUSC EXTINCTIONS AND INTRODUCTIONS

One of the aims of this study was to test accepted hypotheses concerning the chronology of extinction and introduction dates of certain mollusc species into Britain and particularly in Cornwall.

Xerocrassa geyeri

This species is now extinct throughout Britain (Kerney 1999, 184). It is considered to be a cold climate indicator with a restricted modern European distribution (Magnin 1993), although it is capable of surviving in cryptic refugia during major climatic changes (Pfenninger *et al.* 2003). This species has been documented in the very early British post-Glacial period at Upper Halling, Holborough, Kent and Oxted, Surrey (Kerney 1963, 237), at Aylesford, Maidstone, Kent (Burchell and Davis 1957) and at Durrington Walls, Wiltshire (Evans 1971a, 334; Evans 1972, 182).

Only at Gwithian has it been found in later deposits, where previous studies have shown it in early Bronze Age levels. It was identified by Spencer (1974, 1975) at Gwithian (as '*Candidula* cf. *intersecta*' but the correct identification was subsequently confirmed by Milles (1991)). The identification of the specimens in the present study was confirmed by Dr Edmund Gittenberger (pers. com.).

The radiocarbon dating of several specimens of this species in the present study establishes that it was present on the main Gwithian site at least until 2025–1750 cal BC and at Strap Rocks as late as 1635–1500 cal BC. It therefore appears to have survived until towards the end of the early Bronze Age. This was shortly before a period of climatic instability (Figure 168) sufficient to deflate the sand in the trench, and this instability may well have contributed to the final demise of a species that had managed to survive in this small area for several thousand years after it became extinct elsewhere.

Cochlicella acuta

The first appearance of *Cochlicella acuta* in Cornwall is considered to be in the late Bronze Age (Lewis 1968, 315; Lewis 1975; Spencer 1975) although it did not reach Scotland until the Iron Age (Evans 1971b, 59; Evans 2004). Several specimens have been radiocarbon dated in the present study. Those from the old ground surface at western end of Godrevy Towans, where shells were numerous, are dated to 2570–2350 cal BC, placing this species in Cornwall during the late Neolithic. Shells from Strap Rocks are dated to 1745–1610 cal BC, the early Bronze Age. Specimens from a buried soil above the old ground surface at the Towans scarp are dated to 1500–1420 cal BC. The evidence seems conclusive that this species was in Cornwall considerably earlier than previously thought, probably as early as late Neolithic. It could well have been introduced to this area by traders from the Continent, of which there is increasing evidence from at least the early Bronze Age (Van de Noort *et al.* 1999). It may have survived in a local microclimate for at least a thousand years before spreading to other areas of Britain but further research might reveal its presence elsewhere in southern England earlier than the traditionally accepted date.

Helicella itala

Helicella itala is considered to be a native species, i.e. it was established before separation of Britain from the Continent. It is known on the chalk of southern England from the end of the glacial period (Kerney 1999, 182) but the successional appearance of *Helicella* before *Cochlicella* observed by Evans (1979) does not seem to apply in west Cornwall. In every site where bedrock was reached (Gwithian trench, Godrevy Towans and Strap Rocks) *Cochlicella* is found in deeper levels than *Helicella.* This succession was also found at Gwithian by Spencer (1975) and Milles (1991), but not at Atlantic Road, Newquay, Cornwall (Davies 2008, 140), or Brean Down, Somerset, (Bell and Johnson 1990, Fig. 156) where *Helicella* preceded *Cochlicella.* It would seem likely that the environment on the open dunes in west Cornwall was not favourable for *Helicella*, and the conclusion is that this species was a late coloniser of some Cornish dunes, not becoming established until the middle or even late Bronze Age.

Cornu aspersum

This was an infrequent finding during the present study at Gwithian (all sites) although at no location was it found in prehistoric sediments. Its occurrence is consistent with the generally accepted view of its Romano-British introduction (Kerney 1999, 205).

Cernuella virgata

The introduction of this mollusc has been difficult to date as many early records are considered unsatisfactory (Kerney 1966, 10; Evans 1972, 179). In the Godrevy Towans section it did not appear until the early medieval period and in the Gwithian trench probably not until early medieval or later times, although deflation has made this timing unreliable. The accepted view of a Romano-British introduction remains valid.

Candidula intersecta

This species makes its appearance only in very recent times in all the study sites. There is therefore no need to revise the accepted opinion that this a medieval or later introduction (Evans 1972, 179 as *Helicella caperata*; Kerney 1999, 179). It was not present at Strap Rocks, even in the most superficial sediments.

Cecilioides acicula

This mollusc lives a subterranean life, burrowing up to 2m below the ground surface, and its introduction is therefore very difficult to date, there being no reliable Holocene subfossil records (Evans 1972, 168). It was found at Crane Godrevy in levels probably associated with late medieval agricultural activity. It has not previously been recorded on Godrevy Towans (Turk *et al.* 2001, 91). One of the turf specimens was alive when collected proving that the species is extant in the area; it is therefore not possible to determine whether the subfossil specimens are contemporary with the medieval activity or modern.

Potamopyrgus antipodarum

No other introduced species were found in this study apart from *Potamopyrgus antipodarum* in the Red River leet sample; its finding here is consistent with its recorded mid-twentieth century appearance in Cornwall (Forbes and Hanley 1853, 266; Kerney 1999, 36).

CONCLUSIONS

The excavations led by Charles Thomas at Gwithian in the 1950s and 1960s revealed a Bronze Age settlement which was initially established in the Red River valley in the lea of Godrevy Towans around 1800 cal BC. This settlement passed through three major phases of occupation before final abandonment around 900 cal BC; episodes of sand blow resulted in periods when there was considerably reduced activity. There was no further known activity in this area of the valley until the post-Roman period, when an industrial complex was built on a sand dune and occupied from the fifth to eighth centuries AD.

The changes in the palaeoenvironment which relate to this episodic occupation pattern can now be established. When the Bronze Age settlement was first established the valley basin at Gwithian was fen/carr woodland which had probably changed little during the Neolithic period. On the valley sides dune formation with blown sand had commenced in the early Neolithic; there is scant evidence of woodland persisting away from the base of the valley, but it is likely that there was ample timber available locally for that first settlement to be constructed. While there is no evidence of woodland at the west end of Godrevy Towans, to the south of the river estuary at Strap Rocks there was wood/scrub with adjacent marshy land up to the time of the main settlement north of the Red River; the presence of field walls at Strap Rocks implies that there were further settlements, adding to the intensity and extent of known prehistoric activity in the area. The woodland/scrub south of the river did not survive into the middle Bronze Age, and an anthropogenic cause for its loss is likely.

The northern valley sides at Gwithian generally remained stable during the middle and late Bronze Age although sand now claimed the river valley, with loss or significant reduction of marsh land. Variation in settlement activity attests to intermittent sand blow, and this is seen in the 2012 trench where there is a 15cm thick sand horizon separating two stabilisation layers. Farming continued, the original excavations showing evidence of ploughing near the settlement with the current study revealing cattle and ovi-caprid hoof prints in the surface of both stabilisation deposits, much more evident on the upper layer. The field system at Strap Rocks seems to have been overwhelmed with sand by the middle Bronze age, with no evidence of it again being cultivated. At the western end of the Towans there were phases of stability interspersed with phases of sand blow.

By the Iron Age all activity in the valley seems to have ceased. The fields on the valley sides were covered in sand and the valley floor returned to marsh, which persisted until late medieval times. Th siting of the Roman period homestead at Porth Godrevy may have been selected to avoid the unstable conditions in the valley floor. Stability returned to the meadows by the post-Roman period, around AD 400, when industrial activity is evident on a dune to the south-west of the Bronze Age settlement. The farmed area at this time included the meadows, although evidence of this in the area of the 2012 trench is lost perhaps due to deflation of sand.

This area again became unstable around 800 AD, and settlement activity in the northern part of the Red River valley ceased.

There is evidence of Iron Age or Romano-British activity on the summit of Godrevy Towans at Crane Godrevy, but it is not until the late tenth or eleventh centuries that buildings reveal occupation, which lasted to the late sixteenth century. This habitation was surrounded by field systems, shown in the present study to be more extensive to the south-west than previously known.

While there has never been any indication of load mining in the lower reaches of the Red River, it is known that extensive stream mining took place, although never in the section studied in the present work. Geochemistry hints at the possibility of upstream mining for a short period around 1750 BC, but it was not until AD 1050–1100 that there is good evidence of mining residues in the river, and its waters probably turned red. When the river ceased to be known as the Conar Dour and become the Red River is not recorded, but it was before the end of the eighteenth century during the Industrial Revolution (Hamilton Jenkin 1965, 24; Williams 2016), when the waters would certainly have been red. Mining activity increased during the early Industrial Age and continued until very recently when the final pits were closed in the late twentieth century.

The earlier proposition that the lower Red River valley was a marine estuary extending into the low-lying ground between the Bronze Age settlement and the 'post-Roman' sand dune is disproved. Bore hole evidence and the present study suggests it may have extended inland in prehistoric times, but was confined to the centre of the valley and did not reach northwards into the occupation areas. Subsequently the marine inlet became choked by accumulating blown sand until recent times when the lowest reaches have been substantially modified by river re-routing and sand extraction.

The timing of certain mollusc extinctions and introductions can be revised. *Xerocrassa geyeri* is shown to have survived in the Gwithian area until 1635–1500 cal BC, the end of the early Bronze Age, while *Cochlicella acuta* is demonstrated to have been present as early as 2570–2350 cal BC, towards the end of the Neolithic. It appears that there may have been a suitable microclimate at Gwithian which allowed the persistence of one species for many centuries after it became extinct elsewhere in Britain, and for the survival of another species for a period prior to its appearance elsewhere. While *X. geyeri* is generally associated with colder climates and *c. acuta* requires somewhat warmer conditions, both are species of dry open ground, and this area of Cornwall appears to have provided a habitat which somehow met the requirements of both species.

Figure 172. Sunset over Gwithian lighthouse, from Godrevy Towans; June 2016.

Bibliography

Adderley, P., Wilson, C. A., Simpson, I. A. and Davidson, D. A. 2010. Anthropogenic features. In G. Stoops, V. Marcelino and F. Mees (eds) *Interpretation of micromorphological features of soils and regoliths.* Amsterdam: Elsevier, 569–588.

Aitken, M. J. 1994. Optical dating: a non-specialist review. *Quaternary Geochronology (Quaternary Science Reviews)* **13**, 503–508.

Allen, J. R. L. 2000. Holocene coastal lowlands in NW Europe: autocompaction and the uncertain ground. In K. Pye and J. R. L. Allen (eds) *Coastal and estuarine environments: sedimentology, geomorphology and geoarchaeology.* London: Geological Society, 239–252.

Allen, J. R. L. and Thornley, D. M. 2004. Laser granulometry of Holocene estuarine silts: effects of hydrogen peroxide treatment. *The Holocene* **14**, 290–295.

Allen, M. J. 2009. Molluscs. In J. Wright, M. Leivers, R. Seager Smith and C. J. Stevens (eds) *Cambourne New Settlement; Iron Age and Romano-British settlement on the clay uplands of west Cambridgeshire.* Wessex Archaeology Report 23, 187–210.

Allen, M. J. 2017. The geoarchaeology of context: sampling for land snails (on archaeological sites and colluvium). In M. J. Allen (ed.) *Molluscs in archaeology: methods, approaches and applications.* Oxford: Oxbow, 30-47.

Allen, M. J., Sharples, N. and O'Connor, T. (eds) 2009. *Land and people; papers in memory of John G. Evans.* Oxford: Oxbow.

Anderson, R. 2005. An annotated list of the non-marine Mollusca of Britain and Ireland. *Journal of Conchology* **38**, 607–637.

ApSimon, A. M. 2000. Brean Down sand cliff revisited: Pleistocene stratification, new finds and the date of the maritime Bell Beaker. *Proceedings of the University of Bristol Spelaeological Society* **22**, 53–80.

Arkell, W. J. 1943. The Pleistocene rocks at Trebetherick Point, north Cornwall: their interpretation and correlation. *Proceedings of the Geologists' Association* **54**, 141–170.

Bakels, C. C. 1997. The beginnings of manuring in western Europe. *Antiquity* **71**, 442–445.

Ball, D. F. 1964. Loss-on-ignition as an estimate of organic matter and organic carbon in non-calcareous soils. *Journal of Soil Science* **15**, 84–92.

Banerjea, R. Y. 2011. The changing use of internal urban space: a microscopic perspective from mid-Roman timber building 1. In M. G. Fulford and A. S. Clarke (eds) *Silchester: city in transition: the mid-Roman occupation of Insula IX, c. AD125-250/300. A report on excavations undertaken since 1997.* London: Society for the Promotion of Roman Studies. Britannia Monograph 25.

Barber, K. E., Chambers, F. M. and Maddy, D. 2003. Holocene palaeoclimates from peat stratigraphy: macrofossil proxy climate records from three oceanic raised bogs in England and Ireland. *Quaternary Science Reviews* **22**, 521–539.

Barber, K. E., Chambers, F. M., Maddy, D., Stoneman, R. and Brew, J. S. 1994. A sensitive high-resolution record of late Holocene climatic change from a raised bog in northern England. *The Holocene* **4**, 198–205.

Barber, K. E., Maddy, D., Rose, N., Stevenson, A. C., Stoneman, R. and Thompson, R. 2000. Replicated proxy-climate signals over the last 2000 years from two distant UK peat bogs: new evidence for regional palaeoclimate teleconnections. *Quaternary Science Reviews* **19**, 481–487.

Bartlett, R. J. 1988. Manganese redox reactions and organic interaction in soils. In R. D. Graham, R. J. Hannam and N. C. Uren (eds) *Manganese in soils and plants.* Dordercht: Kluwer.

Barton, D. B. 1969. *A history of tin mining and smelting in Cornwall.* Exeter: Wheaton.

Batchelor, C. R., Branch, N. P., Allison, E., Austin, P. A., Bishop, B., Brown, A. G., Elias, S. E., Green, C. P. and Young, D. S. 2014. The timing and causes of the Neolithic elm decline: new evidence from the Lower Thames Valley (London, U.K.). *Environmental Archaeology* **19**, 263–290.

Bell, M. 1981. Seaweed as a prehistoric resource. In D. Brothwell and G. W. Dimbleby (eds) *Environmental aspects of coasts and islands.* Oxford: Oxbow, 117–125.

Bell, M. 1990. *Brean Down excavations 1983-1987.* London: English Heritage.

Bell, M. 2007. *Prehistoric coastal communities: the Mesolithic in western Britain.* York: Council for British Archaeology, CBA Research Report 149.

Bell, M. 2009. Experimental archaeology: changing science agendas and perceptual perspectives. In M. J. Allen, N. Sharples and T. O'Connor (eds) *Land and People: papers in memory of John G. Evans.* Oxford: Oxbow, 31–45.

Bell, M. 2013. Footprint-tracks of people and animals at Redwick. In M. Bell (ed.) *The Bronze Age in the Severn Estuary.* York: Council for British Archaeology, Research Report 172, 144–153.

Bell, M. and Brown, A. 2008. *Southern regional review of geoarchaeology: windblown deposits.* Portsmouth: English Heritage Research Department; report 005/2009.

Bell, M. and Johnson, S. 1990. Non-marine Mollusca. In M. Bell (ed.) *Brean Down Excavations 1983–1987.* London: English Heritage, 246–250.

Bisdom, E. B. A., Stoops, G., Delvigne, J., Curmi, P. and Altemuller, H. J. 1982. Micromorphology of weathering biotite and its secondary products. *Pedologie* **32**, 225–252.

Bishop, A. C., Woolley, A. R. and Hamilton, W. R. 1999. *Minerals, rocks & fossils.* London: Philip.

Blott, S. J. and Pye, K. 2001. Gradistat: a grain size distribution and statistics package for the analysis of unconsolidated sediments. *Earth Surface Processes and Landforms* **26**, 1237–1248.

Bond, G., Kromer, B., Beer, J., Muscheler, R., Evans, M. N., Showers, W., Sharon, H., Lotti-Bond, R., Hajdas, I. and Bonani, G. 2001. Persistent solar influence on north Atlantic climate during the Holocene. *Science* **294**, 2130–2136.

Bond, G., Showers, W., Cheseby, M., Lotti, R., Almasi, P., deMenocal, P., Priore, P., Cullen, H., Hajdas, I. and Bonani, G. 1997. A pervasive millennial-scale cycle in north Atlantic Holocene and Glacial climates. *Science* **278**, 1257–1266.

Booth, G. 1814. *The historical library of Diodorus the Sicilian in fifteen books.* London: Davis.

Borlase, W. 1758. *The natural history of Cornwall.* Oxford.

Borlase, W. 1872. *Nænia Cornubiæ.* London: Longman, Green, Reader and Dyer.

Boycott, A. E. 1934. The habitats of land Mollusca in Britain. *Journal of Ecology* **22**, 1–38.

Boycott, A. E. 1936. The habitats of fresh-water Mollusca in Britain. *Journal of Animal Ecology* **5**, 116–186.

Brady, N. C. and Weill, R. R. 2002. *The nature and properties of soils.* Upper Saddle River: Prentice Hall.

Brain, M. J., Long, A. J., Woodroffe, S. A., Petley, D. N., Milledge, D. G. and Parnell, A. C. 2012. Modelling the effects of sediment compaction on salt marsh reconstructions of recent sea-level rise. *Earth and Planetary Science Letters* **345–348**, 180–193.

Brammer, H. 1971. Coatings in seasonally flooded soils. *Geoderma* **6**, 5–6.

Branch, N., Canti, M. G., Clark, P. and Turney, C. (eds) 2005. *Environmental archaeology; theoretical and practical approaches.* London: Hodder Arnold.

Breuning-Madsen, H., Hoist, M. K., Rasmussen, M. and Elberling, B. 2003. Preservation within long coffins before and after barrow construction. *Journal of Archaeological Science* **30**, 343–350.

Brillouin, L. 1956. *Science and information theory.* New York: Academic Press.

Bristow, C. M. 2011. Some notable Cornish building and decorative stones. *Geoscience in south-west England* **10**, 223–229.

British Geological Survey 1984. *Penzance, sheet 351.358. Solid and drift edition.* Southampton: Ordnance Survey.

British Geological Survey. 2013. *Geology of Britain viewer - borehole scans* [Online]. http://mapapps.bgs.ac.uk/geologyofbritain/home.html?mode=boreholes [Accessed 1 August 2017].

Bronk Ramsey, C., Higham, T. and Leach, P. 2004. Towards high-precision AMS: progress and limitations. *Radiocarbon* **46**, 17–24.

Bronk Ramsey, C., Higham, T., Owen, D. C., Pike, A. W. G. and Hedges, R. E. M. 2002. Radiocarbon dates from the Oxford AMS System: Archaeometry Datelist 31. *Archaeometry* **44 (Supplement 1)**, 1–150.

Brown, A. G. 1997. *Alluvial geoarchaeology: flood plain archaeology and environmental change.* Cambridge: Cambridge University Press.

Buckley, A. 2005. *The story of mining in Cornwall; a world of payable ground.* Fowey: Cornwall Editions.

Buckley, A. 2009. *The Tudor tin industry. The tinners & tin works of Penwith and Kerrier stannary.* Truro: Penhellick Publications.

Buckley, A. 2010. *Dolcoath Mine, a history.* Camborne: The Trevithick Society.

Bull, K. 2011. *Investigation of alluvial deposits in relation to windblown sands sequences at Gwithian, Cornwall: inferences about human activity and environmental change.* Unpublished BSc dissertation, University of Reading.

Bullen, R. A. 1902. Notes on Holocene Mollusca from North Cornwall. *Proceedings of the Malacological Society* **5**, 185–188.

Bullen, R. A. 1912. *Harlyn Bay and the discoveries of its prehistoric remains.* Padstow: Colonel Bellers.

Bullock, P., Fedoroff, N., Jongerius, A., Stoops, G. and Tursina, T. 1985. *Handbook for thin section descripton.* Wolverhampton: Waine Research.

Burchell, J. P. T. and Davis, A. G. 1957. The molluscan fauna of some early post-glacial deposits in north Lincolnshire and Kent. *Journal of Conchology* **24**, 164–170.

Burgess, C. 1976. The Gwithian mould and the forerunners of the south Welsh axes. In C. Burgess and R. Miket (eds) *Settlement and economy in the third and second millennium BC.* Oxford: British Archaeological Reports, British Series 33, 69–79.

Cameron, R. 2016. *Slugs and snails.* London: William Collins.

Cameron, R. A. D. 1978. Interpreting buried land-snail assemblages from archaeological sites - problems and progress. In D. R. Brothwell, K.D. Thomas and J. Clutton-Brock (eds). *Reseach Problems in Zooarchaeology.* London: Institute of Archaeology, occasional publications no. 3, 19–23.

Cameron, R. A. D. 2008. *Keys for the identification of land snails in the British Isles.* Shrewsbury: Field Studies Council.

Cameron, R. A. D. and Morgan-Huws, D. I. 1975. Snail faunas in the early stages of a chalk grassland succession. *Biological Journal of the Linnean Society* **7**, 215–229.

Campbell, S. and Gilbert, A. 1998. The Croyde-Saunton coast. In S. Campbell, C. O. Hunt, J. D. Scourse, D. H. Keen and N. Stephens (eds) *Quaternary of south-west England.* Geological Conservation Review Series **14**; London: Chapman & Hall, 214-224.

Campbell, S., James, H. C. L. and Scourse, J. D. 1999. Godrevy; review of the site. In J. D. Scourse and M. F. A. Furze (eds) *The Quaternary of west Cornwall.* London: Quaternary Research Association, 93–97.

Canti, M. 2007. Deposition and taphonomy of earthwork granules in relation to their interpretive potentional in Quaternary stratigraphy. *Journal of Quaternary Science* **22**, 111-118.

Canti, M. G. 1999. The production of faecal spherulites: animals, environment and taphonomy. *Journal of Archaeological Science* **30**, 135–148.

Canti, M. G. 2003. Earthworm activity and archaeological stratigraphy: a review of products and processes. *Journal of Archaeological Science* **30**, 135–148.

Canti, M. G. and Meddens, F. M. 1998. Mechanical coring as an aid to archaeological projects. *Journal of Field Archaeology* **25**, 97–105.

Carew, R. 1766. *The survey of Cornwall.* Penzance.

Carter, R. W. G. 1988. *Coastal environments.* London: Academic Press.

Childe, V. G. 1931. *Skara Brae. A Pictish village in Orkney.* London: Paul, Trench & Trubner.

Conyers, L. B. 2004. *Ground-penetrating radar for archaeology.* Lanham: AltaMira Press.

Courty, M. A., Goldberg, P. and Macphail, R. 1989. *Soils and micromorphology in archaeology.* Cambridge: Cambridge University Press.

Croudace, I. W., Rindby, A. and Rothwell, R. G. 2006. ITRAX: description and evaluation of a new multi-function X-ray core scanner. In R. G. Rothwell (ed.) *New techniques in sediment core analysis.* London: Geological Society, special publications 267, 51–63.

Cunliffe, B. 2005. *Iron Age communities in Britain.* Abingdon: Routledge.

Cunliffe, B. 2008. *Europe between the oceans; themes and variations: 9000 BC—AD 1000.* New Haven: Yale University Press.

Currant, A. P., Jacobi, R. M. and Rhodes, E. 2006. A new look at the Pleistocene sequence at Brean Down, Somerset, and some observations on the earlier part of the Last Cold Stage in Western Mendip. In C. O. Hunt and S. K. Haslett (eds) *Quaternary of Somerset. Field Guide.* London: Quaternary Research Association, 25–30.

David, A., Linford, N. and Linford, P. 2008. *Geophysical survey in archaeological field evaluation.* Swindon: English Heritage.

Davies, P. 2006. Gwithian 2005 — Land snail assessment. In J. A. Nowakowski, J. Sturgess and A. Lawson Jones (eds) *Gwithian, Cornwall. Report on palaeoenvironmental sampling fieldwork, Scheduled Monument Cornwall 771.* Truro: Historic Environment Service, Cornwall County Council; report 2006R042, 19–21.

Davies, P. 2007. Gwithian 2005 — Land snail assessment. In J. A. Nowakowski (ed.) *Excavations of a Bronze Age landscape and a post-Roman industrial settlement 1953-1961, Gwithian, Cornwall. Assessments of individual key datasets 2003—2006; vol. 1.* Truro: Historic Environment Service, Cornwall County Council; report 2007R017, 64–68.

Davies, P. 2008. *Snails. Archaeology and landscape change.* Oxford: Oxbow.

Davies, P. 2010. Land and freshwater molluscs. In T. P. O'Connor and N. Sykes (eds) *Extinctions and invasions: a social history of British fauna.* Oxford: Windgather, 175–180.
Davies, P., Gale, C. H. and Lees, M. 1996. Quantitative studies of modern wet-ground molluscan faunas from Bossington, Hampshire. *Journal of Biogeography* **23**, 371–377.
de la Vega-Leinert, A.C., Keen, D. H., Jones, R. L., Wells, J. M. and Smith, D. E. 2000. Mid-Holocene environmental changes in the Bay of Skaill, Mainland, Orkney, Scotland: an integrated geomorphological, sedimentological and stratigraphical study. *Journal of Quaternary Science* **15**, 509–528.
Dean, W. E. 1974. Determination of carbonate and organic matter in calcareous sediments and sedimentary rocks by loss on ignition: comparison with other methods. *Journal of Sedimentary Petrology* **44**, 242–248.
Dines, H. G. 1956. *The metalliferous mining regions of south-west England; vol. 1.* London: HMSO.
Doody, P. 2008. *Distribution of sand dunes in Great Britain* [Online]. http://marinespecies.org/introduced/wiki/Distribution_of_Sand_Dunes_in_Great_Britain [Accessed 1 August 2017].
Doody, P. 2009. *Sand dune – country report, Ireland* [Online]. http://marinespecies.org/introduced/wiki/Sand_dune_-_Country_Report,_Ireland [Accessed 1 August 2017].
Duller, G. A. T. 2004. Luminescence dating of Quaternary sediments: recent advances. *Journal of Quaternary Science* **19**, 183–192.
Duller, G. A. T. 2008. *Luminescence dating. Guidelines on using luminescence dating in archaeology.* Swindon: English Heritage.
Durand, N., Curtis Monger, H. and Canti, M. G. 2010. Calcium carbonate features. In G. Stoops, V. Marcelino and F. Mees (eds) *Interpretation of micromorpholocial features of soils and regoliths.* Amsterdam: Elsevier, 143–194.
Dutton, A. and Fasham, P. J. 1994. Prehistoric copper mining on the Great Orme, Llandudno, Gwynedd. *Proceedings of the Prehistoric Society* **60**, 245–286.
Earl, B. 1996. *The Cornish arsenic industry.* Camborne: Penhellick.
Edmonds, R. 1846. On the origin of the sand-hillocks of St. Ives Bay, Whitesand Bay and Mount's Bay. *Transactions of the Royal Geographical Society of Cornwall* **6**, 301–304.
Edmonds, R. 1863. The lost church in the sands of Gwithian in Cornwall. *Archaeologia Cambrensis* **9**, 52–54.
Edwards, R. J. 2006. Mid- to late-Holocene relative sea-level change in southwest Britain and the influence of sediment compaction. *The Holocene* **16**, 575–587.
Ehser, A., Borg, G. and Pernicka, E. 2011. Provenance of the gold of the early Bronze Age Nebra Sky Disk, central Germany: geochemical characterization of natural gold from Cornwall. *European Journal of Mineralogy* **23**, 895–910.
Evans, J. G. 1971a. Durrington Walls: the pre-henge environment. In G. J. Wainwright and I. H. Longworth (eds) *Durrington Walls: excavations 1966–1968.* London: Society of Antiquaries, Report of the Research Committee, no. 29, 329–336.
Evans, J. G. 1971b. Habitat change on the calcareous soils of Britain: the impact of Neolithic man. In D. D. A. Simpson (ed.) *Economy and settlement in Neolithic and early Bronze Age Britain and Europe.* Leicester: University Press, 27–73.
Evans, J. G. 1972. *Land snails in archaeology.* London: Seminar Press.
Evans, J. G. 1979. The palaeo-environment of coastal blown-sand deposits in western and northern Britain. *Scottish Archaeological Forum* **9**, 16–26.
Evans, J. G. 1984. Excavations at Bar Point, St. Mary's, Isles of Scilly, 1979–80. *Cornish Studies* **11**, 7–32.
Evans, J. G. 1991. An approach to the interpretation of dry-ground and wet-ground molluscan taxocenes from central-southern England. In D. R. Harris and K. D. Thomas (eds) *Modelling ecological change.* London: Institute of Archaeology, 75–89.
Evans, J. G. 2004. Land snails as a guide to the environments of wind-blown sand: the case of *Lauria cylindracea* and *Pupilla muscorum*. In A. Gibson and A. Sheridan (eds) *From sickles to circles. Britain and Ireland at the time of Stonehenge.* Stroud: Tempus, 366–379.
Evans, J. G., Davies, P., Mount, R. and Williams, D. 1992. Molluscan taxocenes from Holocene overbank alluvium in central southern England. In S. Needham and M. G. Macklin (eds) *Alluvial Archaeology in Britain.* Oxford: Oxbow, 65–74.

Evans, J. G. and Evans, V. 1995. Land Mollusca. In S. Lobb. Excavation at Crofton causewayed enclosure. *Wiltshire Archaeological and Natural History Magazine* **88**, 22–24.

Everton, A. and Fowler, P. J. 1978. Pre-Roman ard-marks at Lodge Farm, Alveston, Avon: a method of analysis. In H. C. Bowman and P. J. Fowler (eds) *Early land allotment in the British Isles; a survey of recent work.* Oxford: British Archaeological Reports, British Series 48, 179–185.

Fagan, B. 2000. *The Little Ice Age.* New York: Basic Books.

Fitzpatrick, E. A. 1993. *Soil microscopy and micromorphology.* Chichester: Wiley.

Forbes, E. and Hanley, S. 1853. *A history of British Mollusca, and their shells. Vol. IV, Pulmonifera, Gastropoda.* London: van Voorst.

Fowler, P. J. 1962. A native homestead of the Roman period at Porth Godrevy, Gwithian. *Cornish Archaeology* **1**, 17–84.

Fowler, P. J. and Evans, J. G. 1967. Plough-marks, lynchets and early fields. *Antiquity* **41**, 289–301.

Fowler, P. J. and Thomas, A. C. 1962. Arable fields of the pre-Norman period at Gwithian. *Cornish Archaeology* **1**, 61–84.

French, C. 2003. *Geoarchaeology in action.* New York: Routledge.

Gerrard, S. 2000. *The early British tin industry.* Stroud: Tempus.

Gilbertson, D. D., Schwenninger, J.-L., Kemp, R. A. and Rhodes, E. J. 1999. Sand-drift and soil formation along an exposed north Atlantic coastline: 14,000 years of diverse geomorphological, climatic and human impacts. *Journal of Archaeological Science* **26**, 439–469.

Goldberg, P. and Macphail, R. I. 2006. *Practical and theoretical geoarchaeology.* Oxford: Blackwell.

Goode, A. J. J. and Taylor, R. T. 1988. *Geology of the country around Penzance. Memoir for 1:50,000 geological sheets 351 and 358 (England and Wales).* London: HMSO.

Grant, M. J., Waller, M. P. and Groves, J. A. 2011. The *Tilia* decline: vegetation change in lowland Britain during the mid and late Holocene. *Quaternary Science Reviews* **30**, 394–408.

Gregory, R. A., Murphy, E. M., Church, M. J., Edwards, K. J., Guttmann, E. B. and Simpson, D. D. A. 2005. Archaeological evidence for the first Mesolithic occupation of the Western Isles of Scotland. *The Holocene* **15**, 944–950.

Groom, P., Schlee, D., Hughes, G., Crane, P., Ludlow, N. and Murphy, K. 2011. Two early medieval cemeteries in Pembrokeshire: Brownslade Barrow and West Angle Bay. *Archaeologia Cambrensis* **160**, 133–203.

Guttmann, E. B., Simpson, I. A. and Smith, W. 2005. Manuring practices in antiquity: a review of the evidence. In D. N. Smith, M. B. Brickley and W. Smith (eds) *Fertile ground; papers in honour of Susan Limbrey.* Oxford: Oxbow, 68–76.

Hamilton, D., Marshall, P., Roberts, H. M., Bronk Ramsey, C. and Cook, G. 2008. Gwithian: scientific dating. Appendix I in J. A. Nowakowski, H. Quinell, J. Sturgess, C. Thomas and C. Thorpe, Return to Gwithian: shifting the sands of time. *Cornish Archaeology* **46**, 61–70.

Hamilton Jenkin, A. K. 1962. *Mines and miners of Cornwall. III. Around Redruth.* Truro: Truro Bookshop.

Hamilton Jenkin, A. K. 1963. *Mines and miners of Cornwall. V. Hayle, Gwinnear & Gwithian.* Truro: Truro Bookshop.

Hamilton Jenkin, A. K. 1965. *Mines and miners of Cornwall. X. Camborne-Illogan.* Truro: Truro Bookshop.

Hammer, Ø., Harper, D. A. T. and Ryan, P. D. 2001. *PAST: Palaeontological statistics software package for education and data analysis* [Online]. http://palaeo-electronica.org/2001_1/past/issue1_01.htm [Accessed 1 August 2017].

Haslam, W. 1846. The ancient oratories of Cornwall. *The Archaeological Journal* **2**, 225–239.

Hatton, G. 2004. Gwithian: metallurgical debris. *Excavations of a Bronze Age landscape and a post-Roman industrial settlement 1953–1961, Gwithian, Cornwall. Assessments of individual key datasets 2003-2006; vol. 2.* Truro: Historic Environment Service, Cornwall County Council; report 2007R017, 212–216.

Hearne, T. 1769. *The itinerary of John Leland the antiquary.* Oxford.

Heiri, O., Lotter, A. F. and Lemcke, G. 2001. Loss on ignition as a method for estimating organic and carbonate content in sediments: reproducibility and comparability of results. *Journal of Paleolimnology* **25**, 101–110.

Heyworth, A. and Kidson, C. 1982. Sea-level changes in southwest England and Wales. *Proceedings of the Geologists' Association* **93**, 91–111.

Hornblower, S. and Spawforth, A. (eds) 2005. *The Oxford Classical Dictionary.* Oxford: Oxford University Press.

Hosking, K. F. G. and Ong, P. M. 1963–64. The distribution of tin and certain other 'heavy' metals in the superficial portions of the Gwithian/Hayle beach of West Cornwall. *Transactions of the Royal Geological Society of Cornwall* **19**, 351–396.

Howard, P. J. A. and Howard, D. M. 1990. Use of organic carbon and loss-on-ignition to estimate soil organic matter in different soil types and horizons. *Biological and Fertility of Soils* **9**, 306–310.

Howie, F. M. P. and Ealey, P. J. 2010. An appraisal of Quaternary calcium carbonate deposits in Cornwall. *Geoscience in South-West England* **12**, 233–239.

Huddart, D., Roberts, G. and Gonzalez, S. 1999. Holocene human and animal footprints and their relationships with coastal environmental change, Formby Point, NW England. *Quaternary International* **55**, 29–41.

James, H. C. L. 1994. *Late Quaternary coastal landforms and associated sediments of West Cornwall.* Unpublished PhD thesis, University of Reading.

James, H. C. L. 1995. Raised beaches of West Cornwall and their evolving geochronology. *Proceedings of the Ussher Society* **8**, 437–440.

James, H. C. L. 2008. An extension of the well-known Quaternary section at Godrevy, St Ives Bay, West Cornwall: analysis and review. *Geoscience in South-West England* **12**, 55–57.

Johnsen, S. J., Dahl-Jensen, D., Gundestrup, N., Steffensen, J. P., Clausen, H. B., Miller, H., Masson-Delmotte, V., Sveinbjörnsdottir, A. E. and White, J. 2001. Oxygen isotope and palaeotemperature records from six Greenland ice-core stations: Camp Century, Dye-3, GRIP, GISP2, Renland and NorthGRIP. *Journal of Quaternary Science* **16**, 299–307.

Johnstone, E., Macklin, M. G. and Lewin, J. 2006. The development and application of a database of radiocarbon-dated Holocene fluvial deposits in Great Britain. *Catena* **66**, 14–23.

Jones, A. 1998. *Upton and Gwithian Towans: an archaeological assessment.* Truro: Cornwall Archaeological Unit; report 1998R062.

Jones, H. L. 1917. *The geography of Strabo; vol. 1.* London: Heinemann.

Jones, H. L. 1923. *The geography of Strabo; vol. 2.* London: Heinemann.

Juggins, S. 2011. *C2. Software for ecological and palaeoecological data analysis and visualisation (version 1.7.6).* Newcastle upon Tyne: Newcastle University.

Kennard, A. S. and Warren, S. H. 1903. The blown sands and associated deposits of Towan Head, near Newquay, Cornwall. *The Geological Magazine (IV)* **10**, 19–25.

Kerney, M. P. 1963. Late-glacial deposits on the chalk of South-east England. *Philosophical Transactions of the Royal Society of London, Series B, Biological Sciences* **246**, 203–254.

Kerney, M. P. 1966. Snails and man in Britain. *Journal of Conchology* **26**, 3–14.

Kerney, M. P. 1999. *Atlas of land and freshwater molluscs of Britain and Ireland.* Colchester: Harley.

Kerney, M. P. and Cameron, R. A. D. 1979. *A field guide to the land snails of Britain and north-west Europe.* London: Collins.

Kevill, R. 2012. *Were the exposed coastal environments of Gwithian in Cornwall originally woodland environments and what was the role of coastal processes and human activity in changing these environments in the Holocene?* Unpublished BSc dissertation, University of Reading.

Killeen, I., Aldridge, D. and Oliver, G. 2004. *Freshwater bivalves of Britain and Ireland.* Shrewsbury: Field Studies Council.

King, A. 2011. *Strategic stone study. A building stone atlas of Cornwall & Isles of Scilly.* English Heritage.

Kirkham, G. 2005. *Gwithian stream works and Godrevy café, Cornwall. Historic environment assessment and building recording.* Truro: Historic Environment Service, Cornwall County Council; report 2006R042.

Kühn, P., Aguillar, J. and Miedema, R. 2010. Textural pedofeatures and related horizons. In S. G., V. Marcelino and F. Meed (eds) *Interpretation of micromorphological features in soils and regoliths.* Amsterdam: Elsevier, 217-250.

La Motta, V. M. and Schiffer, M. B. 1999. Formation processes of house floor assemblages. In P. M. Allison (ed.) *The archaeology of household activities.* London: Routledge, 19–29.

Lamb, R. C. and Rees, S. E. 1981. Ard cultivation at Sumburgh, Shetland. *Tools and Tillage* **4**, 117-121.

Law, M. 2017. Numerical approaches to land snail palaeoecology. In M. J. Allen (ed.) *Molluscs in archaeology: methods, approaches and applications.* Oxford: Oxbow, 48-63.

Law, M. and Thew, N. 2017. Molluscs from dune-machair system in the Western Isles. In M. J. Allen (ed.) *Molluscs in archaeology: methods, approaches and applications.* Oxford: Oxbow, 82-99.

Lewes, G. 1906. Tin Mining. In W. Page (ed.) *The Victoria County History of the County of Cornwall; vol. 1.* London: Street, 522–570.

Lewis, D. 1992. The sands of time: Cornwall's Hayle to Gwithian Towans. In R. W. G. Carter, T. G. F. Curtis and M. J. Sheehy-Skeffington (eds) *Coastal dunes; geomorphology, ecology and management for conservation.* Rotterdam: Balkema, 463–473.

Lewis, G. 1968. *Polymorphism in the shell characters of certain helicid molluscs, particularly the genus Cochlicella.* Unpublished PhD thesis, University of Oxford.

Lewis, G. 1975. Shell polymorphism in the snail *Cochlicella acuta* (Müller) and some data on its genetics. *Biological Journal of the Linnean Society* **7**, 147–160.

Lewis, H. 2012. *Investigating ancient tillage: an experimental and soil micromorpholoigcal study.* Oxford: British Archaeological Reports, International Series 2388.

Ložek, V. 1990. Molluscs in loess, their paleoecological significance and role in geochronology - principles and methods. *Quaternary International* **7/8**, 71–79.

Macan, T. T. 1977. *A key to the British fresh- and brackish-water gastropods.* Ambleside: Freshwater Biological Association.

Mackenzie, W. S. and Adams, A. E. 1994. *A colour atlas of rocks and minerals in thin section.* New York: Wiley.

Mackenzie, W. S. and Guilford, C. 1980. *Atlas of rock forming minerals.* Harlow: Longman.

Macklin, M. G., Benito, G., Gregory, K. J., Johnstone, E., Lewin, J., Michczyńska, D. J., Soja, R., Starkel, L. and Thorndycraft, V. R. 2006. Past hydrological events reflected in the Holocene fluvial record of Europe. *Catena* **66**, 145–154.

Macphail, R. I. 1998. A reply to Carter and Davidson's "an evaluation of the contribution of soil micromorphology to the study of ancient arable agriculture". *Geoarchaeology* **13**, 549-564.

Macphail, R. I. and Crowther, J. 1996. Short-term, post-burial change in a humic rendzina soil, Overton Down experimental earthwork, Wiltshire, England. *Geoarchaeology: an International Journal* **11**, 95–117.

Magnin, F. 1993. *Trochoidea geyeri* (Soós, 1926) (Pulmonata, Helicidae) in south-eastern France; ecology, biogeography and Quaternary History. *Scripta Geologica; special issue* **2**, 291–295.

Mann, M. E., Zhang, Z., Rutherford, S., Bradley, R. S., Hughes, M. K., Shindell, D., Ammann, C., Faluvegi, G. and Ni, F. 2009. Global signatures and dynamical origins of the Little Ice Age and medieval climate anomaly. *Science* **326**, 1256–1260.

Matthews, W. 1995. Micromorphological characteristics of occupation deposits and microstratigraphic sequences at Abu Salabikh, Southern Iraq. In A. J. Barnham and R. I. Macphail (eds) *Archaeological sediments and soils: analysis, interpretation and management.* London: Insititute of Archaeology, 41–76.

Matthews, W. 2000. Micromorphology of occupation sequences. In R. Matthews and J. N. Postggate (eds) *Contextual analysis of the use of space at two Near Eastern Bronze Age sites.* York: Archaeology Data Service.

Matthews, W. 2010. Geoarchaeology and taphonomy of plant remains and microarchaeological residues in early urban environments in the Ancient Near East. *Quaternary International* **214**, 98–113.

Mattingly, J., Marley, J. and Jones, A. M. 2009. Five gold rings? Early Bronze Age gold lunulae from Cornwall. *Journal of the Royal Institute of Cornwall*, 95–114.

Megaw, J. V. S. 1976. Gwithian, Cornwall: some notes on the evidence for Neolithic and Bronze Age settlement. In C. Burgess and R. Miket (eds) *Settlement and Economy in the Third and Second Millennia B.C.* Oxford: British Archaeological Reports, British Series 33, 51–66.

Megaw, J. V. S., Thomas, A. C. and Wailes, B. 1960–61. The Bronze Age settlement at Gwithian, Cornwall. *Proceedings of the West Cornwall Field Club* **2**, 200–215.

Milles, A. 1991. *The molluscan biostratigraphy and archaeology of Holocene blown-sand in the British Isles.* Unpublished PhD thesis, University of Wales at Cardiff.

Mitchell, P. and Barr, D. 1995. The nature and significance of public exposure to arsenic: a review of its relevance to south west England. *Environmental Geography and Health* **17**, 57–82.

Moore, P. D., Sturgess, J. and Collinson, M. E. 1991. *Pollen analysis.* Oxford: Blackwell.

Morris, R. J., McCartney, M. J., Lockyer, C. and Hoborn, R. 1985. The particulate load of the Red River, St Ives Bay: its geochemical composition and the effect of its discharge plume on the behaviour of a resident wild dolphin. *Marine Pollution Bulletin* **16**, 106–108.

Murphy, E., Gregory, R. and Simpson, D. 2004. Post-Beaker period death and burial at Northton, Isle of Harris, Scotland. *Environmental Archaeology* **9**, 163–171.

Murphy, E. and Simpson, D. 2003. Neolithic Northton: a review of the evidence. In I. Armit, E. Murphy, E. Nelis and D. Simpson (eds) *Neolithic settlement in Ireland and western Britain.* Oxford: Oxbow, 102–111.

Murray, A. S. and Wintle, A. G. 2000. Luminescence dating of quartz using an improved single-aliquot regenerative-dose protocol. *Radiation Measurements* **32**, 57–73.

Murray, A. S. and Wintle, A. G. 2003. The single aliquot regenerative dose protocol: potential for improvements in reliability. *Radiation Measurements* **37**, 377–381.

Musset, S. 2012. *Dune formation: a combination of environmental changes and human activity at Upton Towans, Gwithian, Cornwall.* Unpublished BSc dissertation, University of Reading.

Needham, S. 1996. Chronology and periodisation of the British Bronze Age. *Acta Archaeologica* **67**, 121–140.

Needham, S. 2012. Case and place for the British Chalcolithic. In M. J. Allen, J. Gardiner and A. Sheridan (eds) *Is there a British Chalcolithic? Place, people and polity in the later 3rd millennium.* London: Prehistoric Society, Research Paper 4, 1–16.

Needham, S., Bronk Ramsey, C., Coombs, F., Cartwright, C. and Pettitt, P. 1997. An independent chronology for British Bronze Age metalwork: the results of the Oxford Radiocarbon Accelerator Programme. *Archaeological Journal* **154**, 55–107.

Nowakowski, J. A. 2004. *Archaeology beneath the Towans; excavations at Gwithian, Cornwall 1949–1969. Updated project design; design for assessment, analysis and publication.* Truro: Historic Environment Service, Cornwall County Council.

Nowakowski, J. A. 2007a. *Excavations of a Bronze Age landscape and a post-Roman industrial settlement 1953–1961, Gwithian, Cornwall. Assessment of individual key datasets (2005–2006). Vol. 2.* Truro: Historic Environment Service, Cornwall County Council; report 2007R017.

Nowakowski, J. A. 2007b. *Excavations of a Bronze Age landscape and a post-Roman industrial settlement 1953–1961, Gwithian, Cornwall. Assessment of key datasets (2005–2006). Vol. 1.* Truro: Historic Environment Service, Cornwall County Council; report 2007R017.

Nowakowski, J. A. 2009. Living in the sands – Bronze Age Gwithian, Cornwall, revisited. In M. J. Allen, N. Sharples and T. P. O'Connor (eds) *Land and people; papers in memory of John G. Evans.* Oxford: Oxbow, 115–125.

Nowakowski, J. A. 2011. Telling tales from the roundhouse. Reasearching Bronze Age buildings in Cornwall. In S. Pearce (ed.) *Recent archaeological work in south-western Britain.* Oxford: British Archaeological Reports, British Series 548, 101–120.

Nowakowski, J. A., Quinnell, H., Sturgess, J., Thomas, A., C. and Thorpe, C. 2007. Return to Gwithian: shifting the sands of time. *Cornish Archaeology* **46**, 13–76.

Nowakowski, J. A., Sturgess, J. and Lawson-Jones, A. 2006. *Gwithian, Cornwall. Report on palaeoenvironmental sampling fieldwork, Scheduled Monument Cornwall 771.* Truro: Historic Environment Service, Cornwall County Council; report 2006R042.

Nowakowski, J. A. and Thomas, A. C. 2007. Gwithian - emerging themes and future potential. In J. A. Nowakowski, H. Quinnell, J. Sturgess, C. Thomas and C. Thorpe. Return to Gwithian: shifting the sands of time. *Cornish Archaeology* **46**, 49–61.

O'Connor, T. and Sykes, N. 2010. *Extinctions and invasions: a social history of British fauna.* Oxford: Windgather.

O'Neill, B. H. S. J. 1934. The Roman villa at Magor Farm, Camborne. *Journal of the Royal Insitiute of Cornwall* **24**, 40–69.

PAGES 2k Consortium 2013. Continental-scale temperature variability during the last two millennia. *Nature Geoscience* **6**, 339–364.

Parker, A. G., Goudie, A. S., Anderson, D. E., Robinson, M. A. and Bonsall, C. 2002. A review of the mid-Holocene elm decline in the British Isles. *Progress in Physical Geography* **26**, 1–45.

Peet, R. K. 1975. Relative diversity indices. *Ecology* **56**, 496–498.

Penhallurick, R. D. 1986. *Tin in antiquity; its mining and trade throughout the ancient world with particular reference to Cornwall.* London: The Institute of Metals.

Penhallurick, R. D. 1997. The evidence for prehistoric mining in Cornwall. In P. Budd and D. Gale (eds) *Prehistoric extractive metallurgy in Cornwall.* Truro: Cornwall Archaeological Unit, 23–33.

Pfenninger, M., Posada, D. and Magnin, F. 2003. Evidence for survival of Pleistocene changes in Northern refugia by the land snail *Trochoidea geyeri* (Soós 1926) (Helicellinae, Stylommatophora). *BMC Evolutionary Biology* **3**, 8.

Pirrie, D., Power, M. R., Payne, A., Camm, G. S. and Wheeler, P. D. 2000. Impact of mining on sedimentation; the Camel and Gannel estuaries, Cornwall. *Geoscience in south-west England* **10**, 21–28.

Pirrie, D., Power, M. R., Rollinson, G., Camm, G. S., Hughes, S. H., Butcher, A. R. and Hughes, P. 2003. The spatial distribution and source of arsenic, copper, tin and zinc within the surface sediments of the Fal estuary, Cornwall, UK. *Sedimentology* **50**, 579–595.

Pirrie, D., Power, M. R., Wheeler, P. D., Cundy, A., Bridges, C. and Davey, G. 2002. Geochemical signature of historical mining: Fowey estuary, Cornwall, UK. *Journal of Geochemical Exploration* **76**, 31–43.

Polsue, J. 1868. *A complete parochial history of the county of Cornwall, vol. 2.* Truro: Lake.

Pye, K. and Blott, S. 2004. Particle size analysis of sediments, soils and related particulate materials for forensic purposes using laser granulometry. *Forensic Science International* **144**, 19–27.

Quinnell, H. 2007a. Gwithian: assessment of the prehistoric and Roman period pottery. In J. A. Nowakowski (ed.) *Excavations of a Bronze Age landscape and post-Roman industrial settlement 1953–1961, Cornwall. Assessment of individual key datasets 2003–2006. Vol. 2.* Truro: Cornwall Historic Environment Service; report 2007R017, 95–114.

Quinnell, H. 2007b. The Iron Age and Roman period. In J. A. Nowakowski, H. Quinnell, J. Sturgess, C. Thomas and C. Thorpe. Return to Gwithian: shifting the sands of time. *Cornish Archaeology* **46**, 37–38.

Quinnell, H. 2007c. Neolithic activities on the excavated sites. In J. A. Nowakowski, H. Quinnell, J. Sturgess, C. Thomas and C. Thorpe. Return to Gwithian: shifting the sands of time. *Cornish Archaeology* **46**, 23.

Reille, M. 1992. *Pollen et spores d'Europe et d'Afrique du Nord.* Marseille: Laboratoire de Botanique Historique et Palynologie.

Reimer, P. J., Bard, E., Baylis, A., Beck, J. W., Blackwell, P. G., Bronk Ramsey, C., Buck, C. E., Cheng, H., Edwards, R. L., Friedrich, M., Grootes, P. M., Guilderseon, T. P., Haflidason, H., Hajdas, I., Hatté, C., Heaton, T. J., Hoffmann, D. L., Hogg, A. G., Hughen, K. A., Kaiser, K. F., Kromer, B., Manning, S. W., Miu, M., Reimer, R. W., Richards, D. A., Scott, E. M., Southon, J. R., Staff, R. A., Turney, C. S. M. and van der Plicht, J. 2013. Intcal 13 and marine 13 radiocarbon age calibration curves 0-50,000 years cal BP. *Radiocarbon* **55**, 1869–1887.

Retallack, G. J. 2001. *Soils of the past: an introduction to paleopedology.* Oxford: Blackwell.

Richardson, J. B. 1974. *Metal mining.* London: Allen Lane.

Roberts, A. J. 1987. The later Mesolithic occupation of the Cornish coast at Gwithian: preliminary results. In P. Rowley-Conwy, M. Zvelebil and H. P. Blankholm (eds) *Mesolithic Northwest Europe: recent trends.* Sheffield: Department of Archaeology and Prehistory, Sheffield University, 131–137.

Roberts, H. M. 2006. Optically stimulated luminescence (OSL) dating of sands from a Bronze Age archaeological site at Gwithian, West Cornwall, U.K. In J. A. Nowakowski, J. Sturgess and A. Lawson Jones (eds) *Gwithian, Cornwall. Report on palaeoenvironmental sampling fieldwork, Scheduled Monument Cornwall 771.* Truro: Historic Environment Service, Cornwall County Council; report 2006R042, 63–75.

Roberts, H. M. 2008. The development and application of luminescence dating to loess deposits: a perspective on the past, present and future. *Boreas* **37**, 483–507.

Robson, J. 1943–44. The recent geology of Cornwall. *Transactions of the Royal Geographical Society of Cornwall* **17**, 132–163.

Rogers, W. 1909. The raised beaches and Head of the Cornish coast. *Transactions of the Royal Geological Society of Cornwall* **13**, 352–384.

Rogers, W. 1910–11. A kitchen midden on Godrevy Towans. *Journal of the Royal Institution of Cornwall* **18**, 238–240.

Rowe, J. 1953. *Cornwall in the age of the Industrial Revolution.* Liverpool: University Press.

Rowell, D. L. 1994. *Soil science: methods and application.* Harlow: Longman.

Santisteban, J. I., Mediavilla, R., López-Pamo, E., Dabrio, C. J., Zapata, M. B. R., Varcia, M. J. G., Castanño, S. and Martinez-Alfara, P. E. 2004. Loss on ignition: a qualitative or quantitative method for organic matter and carbonate mineral in sediments? *Journal of Paleolimnology* **32**, 287–99.

Saye, S. E. 2003. *Morphology and sedimentology of coastal sand dune systems in England and Wales.* Unpublished PhD thesis, University of London.

Schiffer, M. B. 1996. *Formation processes of the archaeological record.* Salt Lake City: University of Utah Press.

Schmidt, P. K. and Burgess, C. M. 1981. *The axes of Scotland and Northern England.* München: Beck'sche Verlagsbuchhandlung.

Shail, R. K. 1989. Gramscatho-Mylor facies relationships; Hayle, south Cornwall. *Proceedings of the Ussher Society* **7**, 125–130.

Shannon, C. E. and Weaver, W. 1949. *The mathematical theory of communication.* Urbana: University of Illionois Press.

Sharpe, A. 1997. Smoke but no fire, or the mysterious case of the missing miners. In P. Budd and D. Gale (eds) *Prehistoric extractive metallurgy in Cornwall.* Truro: Cornwall Archaeological Unit, 35–9.

Shennan, I. and Horton, B. 2002. Holocene land- and sea-level changes in Great Britain. *Journal of Quaternary Science* **17**, 511–526.

Shepherd, I. A. G. and Tuckwell, A. N. 1976–77. Traces of Beaker-period cultivation at Rosinish, Benbecula. *Proceedings of the Society of Antiquaries of Scotland* **108**, 108–113.

Simpson, D. 1966. A Neolithic settlement in the Outer Hebrides. *Antiquity* **40**, 137–139.

Simpson, I. 1998. Land management at Tofts Ness, Sanday, Orkney: the evidence of thin section micromorphology. In C. M. Mills and G. Coles (eds) *Life on the edge: human settlement and marginality.* Oxford: Oxbow Monograph 100, 91–98.

Smith, H. 1994. *Middening in the Outer Hebrides: an ethnoarchaeological investigation.* Unpublished PhD thesis, University of Sheffield.

Smith, H. 2012. The ethnohistory of Hebridean agriculture. In M. Parker Pearson (ed.) *From machair to mountains. Archaeological survey and excavation in South Uist.* Oxford: Oxbow, 379–400.

Sparks, B. W. 1959–60. The ecological interpretation of Quaternary non-marine Mollusca. *Proceedings of the Linnean Society of London* **172**, 71–80.

Sparks, B. W. 1960. Land Mollusca. In J. Alexander, P. C. Ozanne and A. Ozanne. Report on the investigation of a round barrow on Arreton Down, Isle of Wight. *Proceedings of the Prehistoric Society* **26**, 299–301.

Sparks, B. W. 1961. The post-glacial deposits at Apethorpe, Northamptonshire. *Proceedings of the Malacological Society* **34**, 302–315.

Spencer, P. J. 1974. *Environmental change in the coastal sand dune belt of the British Isles.* Unpublished BSc dissertation, University of Wales at Cardiff.

Spencer, P. J. 1975. Habitat change in coastal sand-dune areas: the molluscan evidence. In J. G. Evans, S. Limbrey and H. Cleere (eds) *The effect of man on the landscape: the Highland zone.* York: Council for British Archaeology, 96–103.

Stace, C. 2005. *New flora of the British Isles.* Cambridge: Cambridge University Press.

Stephens, F. J. 1899. Alluvial deposits in the lower portion of the Red River valley, near Camborne. *Transactions of the Royal Geological Society of Cornwall* **12**, 324–335.

Stephens, F. J. 1899–1900. Alluvial deposits in the lower portion of the Red River valley, near Camborne. *Journal of the Royal Institution of Cornwall* **14**, 51–52.

Stevenson, J. B. 1984. The excavation of a hut circle at Cùl a'Bhaile, Jura. *Proceedings of the Society of Antiquaries of Scotland* **114**, 127–160.

Stoops, G. 2003. *Guidelines for analysis and description of soil thin sections.* Madison: Soil Science Society of America.

Sturgess, J. 2004. *The Gwithian project; summaries and evaluations of enumerated sites.* Truro: Historic Environment Service, Cornwall County Council; report 2004R069.

Sturgess, J. 2007a. Bronze Age Gwithian. In J. A. Nowakowski, H. Quinnell, J. Sturgess, C. Thomas and C. Thorpe. Return to Gwithian: shifting the sands of time. *Cornish Archaeology* **44**, 23–33.

Sturgess, J. 2007b. Post-Roman Gwithian. In J. A. Nowakowski, H. Quinnell, J. Sturgess, C. Thomas and C. Thorpe. Return to Gwithian: shifting the sands of time. *Cornish Archaeology* **44**, 38–44.

Sturgess, J. and Lawson Jones, A. 2006a. *Bronze Age Gwithian revisited. Archaeological excavations between 1956 and 1961 in Cornwall.* Truro: Historic Environment Service, Cornwall County Council; report 2006R067, 2 vols.

Sturgess, J. and Lawson Jones, A. 2006b. *Crane Godrevy revisited. Archaeological excavation on the enclosure ditch between 1956 and 1969 in Cornwall.* Truro: Historic Environment Service, Cornwall County Council; report 2006R066.

Sturgess, J. and Lawson Jones, A. 2006c. *Post-Roman Gwithian revisited. Archaeological excavations between 1953 and 1958 in Cornwall.* Truro: Historic Environment Service, Cornwall County Council; report 2006R072, 2 vols.

Thomas, A. C. 1954–55. Excavations at Gwithian, 1955. *Proceedings of the West Cornwall Field Club* **1**, 122–123.

Thomas, A. C. 1957–58. The Palaeolithic and Mesolithic periods in Cornwall. *Proceedings of the West Cornwall Field Club* **2**, 5–12.

Thomas, A. C. 1958. *Gwithian; ten years' work (1949–1958).* Gwithian: West Cornwall Field Club.

Thomas, A. C. 1964a. *Gwithian. Notes on the church, parish and St. Gothian's Chapel.* Redruth: Earle.

Thomas, A. C. 1964b. Minor sites in the Gwithian area (Iron Age to recent times). *Cornish Archaeology* **3**, 37–62.

Thomas, A. C. 1969. Excavations at Crane Godrevy, Gwithian, 1969; interim report. *Cornish Archaeology* **8**, 84–88.
Thomas, A. C. 1970. Bronze Age spade marks at Gwithian, Cornwall. In A. Gailey and A. Fenton (eds) *The spade in northern and Atlantic Europe.* Belfast: Ulster Folk Museum, Queen's University, 10–17.
Thomas, A. C. 2007. A detailed chronology of Gwithian. In J. A. Nowakowski, H. Quninell, J. Sturgess, C. Thomas and C. Thorpe. Return to Gwithian: shifting the sands of time. *Cornish Archaeology* **46**, 21–23.
Thomas, C. 1871. *Mining fields of the west: being a practical exposition of the principal mines and mining districts of Cornwall and Devon.* St Austell: Warne.
Thomas, K. D. 1985. Land snail analysis in archaeology: theory and practice. In N. R. J. Fieller, D. D. Gilbertson and N. G. A. Ralph (eds) *Palaeobiological investigations; research design, methods and data analysis.* Oxford: British Archaeological Reports, International Series 258, 131–155.
Thomas, W. 1913. Losses in the treatment of Cornish tin ores. *Transactions of the Cornish Institute of Mining, Mechanical and Metallurgical Engineers* **1**, 56–74.
Thomson, J. O. 1965. *History of ancient geography.* New York: Biblo and Tannen.
Thorndycraft, V. R., Pirrie, D. and Brown, A. G. 2004. Alluvial records of medieval and prehistoric tin mining on Dartmoor, southwest England. *Geoarchaeology: an International Journal* **19**, 220–236.
Thorpe, C. and Thomas, A. C. 2008. Post-Roman material culture. In J. A. Nowakowski, H. Quinnell, J. Sturgess, C. Thomas and C. Thorpe. Return to Gwithian: shifting the sands of time. *Cornish Archaeology* **46**, 44–48.
Timberlake, S. 2003. *Excavations on Copa Hill, Cwmystwyth (1986–1999). An early Bronze Age copper mine within the uplands of Central Wales.* Oxford: British Archaeological Reports, British Series 348.
Timberlake, S. 2009. Copper mining and production at the beginning of the British Bronze Age. In P. Clark (ed.) *Bronze Age connections; cultural contact in prehistoric Europe.* Oxford: Oxbow, 94–121.
Turk, S. M., Meredith, H. M. and Holyoak, G. A. 2001. *Atlas of the land and freshwater molluscs of Cornwall and the Isles of Scilly.* Truro: Environmental Records Centre for Cornwall and the Isles of Scilly.
Van de Noort, R., Middleton, R., Foxon, A. and Bayliss, A. 1999. The 'Kilnsea-boat', and some implications from the discovery of England's oldest plank boat remains. *Antiquity* **73**, 131–135.
Van Geel, B., Buurman, J. and Waterbolk, H. T. 1996. Archaeological and palaeoecological indications of an abrupt climate change in the Netherlands, and evidence for climatological teleconnections around 2650 BP. *Journal of Quaternary Science* **11**, 451–460.
Vaughan, M. P. 1976. *Environmental change in areas of blown sand on the western coasts of the British Isles.* Unpublished BSc dissertation, University of Wales at Cardiff.
Walker, T. 2017. Molluscs and the palaeo-environment of coastal blown sand and dunes. In M. J. Allen (ed.) *Molluscs in archaeology: methods, approaches and applications.* Oxford: Oxbow, 65–81.
Walker, T. M. 2010. *Molluscs, environment and archaeology at Godrevy Towans, Cornwall.* Unpublished BSc dissertation, University of Reading.
Walker, T. M. 2014. *Shifting sand: the palaeoenvironment and archaeology of blown sand in Cornwall.* Unpublished PhD thesis, University of Reading.
Walker, Y. 2008. *Analysis of molluscs from Brean Down: the reconstruction of a Bronze Age environmental sequence.* Unpublished BSc dissertation, University of Reading.
Waller, M. P. 1994. Paludification and pollen representation: the influence of wetland size on *Tilia* representation in pollen diagrams. *The Holocene* **4**, 430–434.
Wang, T., Surge, D. and Mithen, S. 2012. Seasonal temperature variability of the Neoglacial (3300–2500 BP) and Roman Warm Period (2500–1600 BP) reconstructed from oxygen isotope ratios of limpet shells (*Patella vulgata*), Northwest Scotland. *Palaeogeography, Palaeoclimatology, Palaeoecology* **317–318**, 104–113.
Webster, C. J. (ed.) 2007. *The archaeology of south west England. South West Archaeological Research Framework. Resource asssesment and research agenda.* Taunton: Somerset Heritage Service, Somerset County Council.
Wessex Archaeology 2002. *Gwithian sand pit, Hayle, Cornwall. Archaeological baseline survey. Auger survey results and archaeological significance.* Salisbury: Wessex Archaeology; report 520.19.1.
Whimster, R. 1977. Harlyn Bay reconsidered: the excavations of 1900–1905 in the light of recent work. *Cornish Archaeology* **16**, 61–88.
Williams, A. and Martin, G. H. (eds) 2003. *Domesday Book. A complete translation.* London: Penguin.
Williams, P. 2016. *The Red River.* Cornwall: Godrevy Press.

Woodward, B. B. 1908. Notes on the drift and underlying deposits at Newquay, Cornwall. *The Geological Magazine (V)* **5**, 10–18, 80–87.

Xu, J. and Thornton, I. 1985. Arsenic in garden soils and vegetable crops in Cornwall, England: implications for human health. *Environmental Geography and Health* **7**, 131–133.

Yim, W. W.-S. 1981. Geochemical investigations on fluvial sediments contaminated by tin-mine tailings, Cornwall, England. *Environmental Archaeology* **3**, 245–256.

Younger, P. L. 2002. Mine water pollution from Kernow to Kwazulu-Natal: geochemical remedial options and their selection in practice. *Geoscience in south-west England* **10**, 255–266.

Index

Molluscs

(only included when named in text; molluscs named in the tables are not indexed)

Plants and other botanical species